TRIBAL
SCIENCE

TRIBAL SCIENCE

Brains, Beliefs, and Bad Ideas

MIKE McRAE

Prometheus Books

59 John Glenn Drive
Amherst, New York 14228–2119

Published 2012 by Prometheus Books

Cover design by Jacqueline Nasso Cooke

Inquiries should be addressed to
Prometheus Books
59 John Glenn Drive
Amherst, New York 14228–2119
VOICE: 716–691–0133
FAX: 716–691–0137
WWW.PROMETHEUSBOOKS.COM

16 15 14 13 12   5 4 3 2 1

Library of Congress Cataloging-in-Publication Data

McRae, Mike, 1977–
    Tribal science : brains, beliefs, and bad ideas / Mike McCrae.
        p. cm.
    Includes bibliographical references and index.
    ISBN 978–1–61614–583–5 (pbk. : alk. paper)
    ISBN 978–1–61614–584–2 (ebook)
    1. Science—Social aspects—History. 2. Human behavior. 3. Social psychology.
I. Title.

Q175.46.M37 2012
303.48'309—dc23

                                                                    2012000380
        Printed in the United States of America on acid-free paper.

For Asher

CONTENTS

INTRODUCTION

I spent Christmas 2004 at a seaside town in the English region of Cornwall. The damp British winter had kept the beaches rather quiet, providing solitude perfect for losing myself in thought. One memory that remains with me is a late afternoon stroll by the cliffs, the fading sun bleeding into the landscape's muted colors while a flock of tiny birds rode the air currents that spilled invisible and silent against the rocks. There was a distinctive edge to the flock's formation, as if it was a giant amoeba that moved by stretching out an immense arm and dragging itself across the sky in a panic before concentrating into a teardrop. It wasn't a clump formed by hundreds of individuals—from my position, it operated as if it was a single animal with one mind.

I've always been fascinated by the behavior of swarms, hives, flocks, schools, colonies and communities. To think the anatomy of a bee is fundamentally suicidal, sacrificing the life of an individual for the sake of saving its colony. Social insects with nervous systems so simple are capable of great feats of computation in foraging and construction, providing us with novel ways to model communication networks for optimal traffic.

Yet when I take a moment to consider it, my fascination with collective behavior and swarm intelligence comes down to a rather arbitrary delineation of units. I don't think of my body's cells as individual entities, myself as an ecosystem or metropolis of single-celled organisms. In some ways, however, I can consider myself as more or less a super-colony of membrane-enclosed bodies cooperating as discrete components. Somewhere in my brain a line is drawn around the outside of my skin, marking that boundary as a defining limit of me as "one" rather than "many."

We readily consider a single bacterial cell to be an individual, while a colony of identical bacteria isn't. Only once cells adopt specific roles, with some forming a protective boundary while others adopt functions responsible for digestion, synthesis or reproduction, does the collective begin to take on the attributes of an individual. On the other hand, bees with physiological and behavioral differences within their hive don't make us view the hive as an individual. The fact a single insect can fly away forces us to define it as an independent organism. But the cells of a sea sponge can migrate throughout its body and change jobs, and entire sections of sponge can break away and settle elsewhere to grow. Is a single sponge a colony, or a discrete animal?

It might seem rather pedantic to give this much thought, but when it comes to understanding behavior—especially that of humans—the innate biases we hold about the concept of social thinking and respect for our state of individuality can lead us astray. We're so hung up on defending free will and theory of mind that the idea of being influenced by the collective to any significant extent feels somewhat heretical, especially in cultures that celebrate the rights of individual freedom. It's easy to see hive-like behavior in others, of course . . . but never ourselves.

But to what extent are we isolated from other minds? How much control do we have over our ideas, our thoughts and our beliefs?

We are all products of our own collectives, our own tribes. That's not to say we're at the complete mercy of external cognitive forces.

But how we behave as individuals can't easily be teased out of the context of a collective either. We are tribal animals.

In today's post-industrial, westernized world, many of us belong to dozens of tribes that each exerts pressure on how we think and make decisions. I was born into a tiny tribe that consisted of my parents, and was later joined by a brother and sister. I found a place among the members of another tribe with friends during primary school and then again in high school. Work colleagues have welcomed me into their tribes here and overseas, as have fellow travelers I met while living in London and touring Europe. I have been an authority for the tribes of students I taught; a faceless member of the digital tribes of online forums and bulletin boards . . . each tribe different in values and beliefs, influencing how I think and act.

Somehow, out of that mess of shared stories, in-jokes, language variations and values, our vision of the world is formed. That human collage can make the difference between accepting that the Earth was made in seven days by the whim of God or over millions of years by impersonal laws of nature. Our tribal knowledge influences our trust in the scientist, the priest, the politician or the celebrity who tells us the globe is warming, vaccinations are unsafe, mobile phones give you cancer or Bill Clinton is really a reptilian alien in disguise.

This book is about ideas and how they develop: not just inside one brain, but from numerous brains working together as a collective. I look at how the brain evolved to help us survive an ever-changing world as cohesive communities, and how its functions bind us together in tribes using meaningful stories about the world to allow us to better face adversity together rather than on our own.

Our nervous system's ability to transfer and receive information with others in our community is what allows us to determine who is friend and who is stranger, what is safe and what is dangerous. Yet somewhere in that process of exchange, a culture has emerged that values a critical comparison of ideas in light of diverse observations. This methodology—what we describe today as science—allows us to prioritize ideas that are likely to be useful in making predictions. Armed with this future sight we have imagined possibilities and made them a reality. Driven by a desire for comfort and an aversion to pain and loss, we can invent ways to change our environment that no other organism is capable of.

For its pragmatic benefits, science still relies on the same cognitive skills of our ancestors. It remains a social endeavor, residing not in any single brain but somewhere in the process of critical debate and strained consensus. The history of science in the western world might be told by way of solitary names, of Galileos, Newtons and Einsteins, but their discoveries were tribal efforts, their contributions merely the critical tweaks that allowed an idea pregnant with possibility to give birth.

Understanding how science operates as a social phenomenon will be vital as our world becomes increasingly dependent on its products. Teaching science in the future will need to be less a task of presenting ideas as a list of facts, and more a facilitation of critical epistemology, where education is delivered as a set of skills and values that work with our social brains rather than in spite of them. For individuals, knowing the difference between a useful belief and a figment of wishful thinking could make the difference between living and dying. As a species, the choice is one which will lead to the suffering of our future generations or their prosperity in a changing world.

Chapter 1

THE STORYTELLING MONKEY

Why do we see faces in clouds?

T he average human brain contains roughly 100 billion neurons, with a typical cell sharing between 1,000 and 10,000 synapses. It can weigh somewhere around 1.3 kilograms in the average adult and account for about 20 percent of the energy you use while at rest. On its own, it's quite impressive, allowing you to drive, sing, write shopping lists, make pancakes, follow movie plot lines and calculate a weekly budget. Yet this is only the tip of the iceberg—what makes it truly a wonder of nature is how it works collectively with other brains.

HARVEY'S SPECIAL BRAIN

On the morning of April 18, 1955, Albert Einstein's brain came out of its skull. Fortunately for the famous physicist, he no longer had much use for it. Seventy-six years, four weeks and six days after his birth in the German city of Ulm on the River Danube, his abdominal aorta ruptured, depriving his brain of the oxygen it so desperately needed to maintain his body's functioning.

While he was alive, Einstein had developed something of a personal attachment to his brain and found the thought of it sitting on a collector's shelf less than appealing. He expressed to his family that his remains should be cremated and the ashes scattered to the four winds. Keeping in mind that Galileo's mummified fingers are still being passed around like a macabre show-and-tell centuries after his death, it's easy to sympathize with Einstein's anxiety over the eventual fate of his body parts.

The pathologist on duty at Princeton Hospital—a 42-year-old physician by the name of Dr Thomas Harvey—was a last-minute replacement. For whatever reason, he decided that the scientist's autopsy called for the removal of the physicist's brain, perhaps for analysis. Or, just maybe, faced with the 1.2 kilogram, gray-and-white ball of wrinkly jelly that was responsible for deducing some of the most revolutionary laws of physics in history, the thought of cutting it into sections and preserving it as a memento was just too tempting. For good measure, the physicist's eyes were also removed and given to Einstein's ophthalmologist, presumably as a gift. While poor old Einstein may not have approved, it's alleged that his son Hans Albert agreed (if rather begrudgingly) that the doctor could hold onto the brain for scientific purposes.

Not everybody was content with this arrangement. Princeton Hospital's director was so displeased with the pathologist's audacious decision that he promptly gave him the boot. Dr Harvey took Einstein's brain and moved to Philadelphia, where over several months it was diced into 170 cubes and embedded in a jelly-like substance called *celloidin* to further preserve it, before being thinly sliced, stained and mounted on microscope slides. With his trophy safe, Dr Harvey carried it around with him wherever he went, mailing off sections to any researcher who expressed curiosity.

If Thomas Harvey had hoped to one day tease out the secrets of Einstein's remarkable intelligence and secure fame and fortune, he was poorly mistaken. For one thing, the man wasn't a neurologist. And while he might have been inspired by the recent discovery that Lenin's brain contained relatively large cortical pyramidal cells (a type of nerve cell), the pathologist didn't seem to have any specific plans for his new acquisition.

Over the years, after Harvey's marriage fell apart and he lost his medical license following a failed exam in Missouri, he decided enough was enough. In 1998, the doctor drove across the United States from New Jersey to California in the company of writer Michael Paterniti. He intended to meet with Evelyn Einstein, Hans Albert's adopted daughter, perhaps in hope of clearing the air over past grievances. In the back of the car the remains of Einstein's brain sat immersed in a transparent liquid inside a plastic Tupperware container.

Sadly, neither the doctor's story nor that of the famous brain would end quite so neatly. The meeting with Evelyn was uncomfortable and didn't result in the return of her grandfather's remains or in Harvey's absolution. Shortly after the road trip he sent his prize back to the pathology lab in Princeton Hospital where it all began.

Out of all of Albert Einstein's organs, it was his brain that was ultimately deemed responsible for the greatest revolution in physics since Isaac Newton's work on gravity. This chunk of gray tissue was held responsible for changing the world. What made it so special? Why that brain and not somebody else's? This is the key question that numerous scientists have pondered over the decades, prompting them to seek out tiny pieces of Einstein's cerebral tissue to analyze.

Whether the pathologist Thomas Harvey was motivated by a desire for knowledge or just seeking a nifty souvenir, experienced neurologists have probed those blocks of neurons for their secrets in search of some particular feature that could help explain such remarkable thoughts. In spite of a few tantalizing clues, Einstein's brain has really told us very little about what makes one human more intelligent than another.

BRAINS—NOT JUST FOR ZOMBIES

Thinking of the brain is to think about thinking. It's associated with all the things that make you who you are. But to the Egyptians it was the human heart that contained the personality. The Greeks and Romans thought the brain was a radiator for cooling the blood or, according to the ancient Roman physician Galen, for producing semen. Given its pale coloring and gelatinous consistency, it's not difficult to see how such a connection could be drawn. To be fair, the brain was also considered to house the spirit, or "animal soul," responsible for controlling a person's involuntary movements.

Yet other ancient biologists managed to at least connect the brain with control and cognition. The 11th-century Persian polymath Abu Ali Sena—or "Avicenna" as he was known in Europe—concluded it was the fluid-filled gaps in the brain called "ventricles" that performed the mental work; imagination swished around in the front one, memories accumulated in the back and reason was somewhere in between.

Today, it's easy to think of Bob Smith having a heart transplant and still being Bob Smith. Swap Bob's brain with Jack's, however, and it would be a different story. So at some point in history, it became clear that the stuff between our ears is responsible for our thoughts, our memories and our desire to collect peanuts shaped like Hollywood movie stars. Our brain defines who we are and how we behave. However, it's difficult reconciling our sense of self with a mechanical process arising from tissues, cells and chemical reactions. How can networks of chemical circuits result in that sensation of free will or love, or our ability to dream about things we've never seen?

Cutting up bodies to take a peek inside has long been considered taboo, forcing early anatomists to make educated guesses on what our organs do. Occasionally word would spread of an injury that might produce obvious trauma or changes in health or behavior, prompting speculation on the specific roles of particular body parts. Animal dissections might also have provided a few tantalizing clues. But without taking out a person's heart, liver, kidneys or brain and having a close look, nobody could be certain of what made human organs so different from those of a dog or monkey.

It wasn't until the Renaissance that attitudes toward anatomical dissections began to relax and physicians were permitted to dissect the bodies of those who had died in their care. Still, autopsies on healthy bodies were less common, restricted to criminals or those who gave their remains up for science.

In 1664, the English anatomist Thomas Willis published *Anatomy of the Brain*. Taking hints from fellow Brit William Harvey—famed for mapping the flow of blood through the circulatory system—Willis tracked the movement of fluid through the cavities of the nervous system by injecting them with ink. He determined that the brain's cavernous ventricles were channels for circulating spinal fluid, dismissing any notion of souls and spirits governing the workings of the brain.

It mightn't seem so revolutionary to us today, but thinking of the brain in physical, mechanical terms like the heart or lungs was a rather significant deviation from tradition. It had become possible to study the brain like any other organ—if the brain broke, a person apparently wouldn't function quite as they had previously. The science of neurology was born, and with it the quest to understand how this knotted mass of nerve cells functioned to make us remember, think and act as we do.

Naturally, to assign roles to the various regions of the brain it pays to find unique individuals and—on their death—open their skull to see what makes them so different. Even today, patients who present various disorders play a vital role in locating neurological functions. Fortunately, with machines that can scan tissue while the patient is still alive, dissections are no longer as necessary as they once were.

The tale of the New Hampshire railroad worker Phineas Gage exemplified the newfound fascination people held for mapping the functions of the brain.

Early in the 19th century, the method of choice for clearing a path for new railroads wouldn't exactly meet modern safety standards. Once a hole was bored into the obstructing rock it was filled with explosives and a handful of sand. To get the biggest bang out of the delicate, highly combustible powder, the worker had to pack it in tight by hammering it with a giant iron rod called a "tamping iron.'

One day in September 1848, Phineas Gage's enthusiastic pounding —and perhaps his absent-minded decision to not use any sand—set off the charge prematurely and launched the iron bar up into his left cheek, through his eye socket and out of the top of his head. New Hampshire railroad workers in the 19th century were bred tough, it seems, as not only did Phineas survive, he managed to sit upright for the entire ride to the inn where he lived. He was working back at his old job again within eight months.

Gage would live a healthy existence for another 12 years after this unfortunate event, eventually dying of seizures caused by the trauma. Yet during this period he became something of a celebrity and traveled around the country with the offending tamping iron in hand, even appearing in PT Barnum's American Museum as an attraction.

If the average person on the street was interested in gawking at Gage's scars and hearing of his miraculous survival, scientists were more interested in possible changes to his personality and other mental functions. Accounts describe how friends noticed Phineas was no longer himself, while psychologists reported a range of psychological abnormalities and changes. On the other hand, a doctor who knew him well found he had no noticeable mental impairment, not to mention that he was of sound enough mind to be employed as a coach driver later in his life. Given little had been recorded of his life before the accident, it's difficult to know precisely how Phineas had really changed as a result of his brain trauma, if at all. It's also impossible to show which aspects of his personality might have come from emotional trauma or lifestyle changes and which were from the missing pieces of his frontal lobe.

Overall, the study of Phineas Gage contributed surprisingly little to the field of neurology, in spite of his fame. However, it served as an illustration to the growing interest scientists had in mapping the brain. It was a new anatomical frontier full of mountains and valleys with unknown functions. And hidden in there somewhere were features that undoubtedly distinguished human reason from animal instinct.

BROCA'S BIG BRAIN

Like explorers of old, anatomists who discovered new territory in the body were occasionally fortunate enough to have it named after them. Paul Broca—a French physician and anthropologist—was one such fellow.

A number of Broca's patients exhibited particular difficulties in producing speech: a condition called "aphasia." One example was a poor young gentleman who could only say the word "tan" over and over. On his death his autopsy revealed a lesion in the left half of his brain, presumably caused by syphilis. The area has become known as "Broca's area" and is still regarded today as playing a key role in producing speech.

Broca might have been right about the neurology of speech, but some of his other ideas were a little more controversial. In 1861, at a meeting of the Anthropological Society of Paris, a colleague named Louis Pierre Gratiolet presented a rather shocking paper challenging a common belief among anthropologists of the day—that big brains equaled greater intelligence.

Of course, Broca was a firm believer that this was the case, so he wasn't too pleased with Gratiolet's argument.

"In general, the brain is larger in mature adults than in the elderly, in men than in women, in eminent men than in men of mediocre talent, in superior races than in inferior races," Broca asserted defensively at the meeting. Basically, if you were a woman, a worker, black or old, you naturally weren't as intelligent as a young, white, upper-class gentleman. All because your brain was smaller.

"Other things being equal," he went on to say, "there is a remarkable relationship between the development of intelligence and the volume of the brain."

It'd be unfair to pick on Broca—Gratiolet's statement that there was no relationship between a person's brain size and their intelligence was not only running against the cultural grain of the time, it also conflicted with the accepted science. Many anthropologists had spent years measuring skulls, conducting intelligence tests and confirming that Caucasian males were vastly smarter—and therefore superior—to all others. Could they all possibly be wrong?

Well, for the most part, yes, quite easily. Some people do, of course, have slightly larger brains. But they also tend to have bigger bodies. More muscles, skin and bigger organs require more brain tissue to control it all. Women's brains are often smaller because they tend to be more petite than your average bloke. So there has to be more to intelligence than "big brain = smart/small brain = stupid."

Studying intelligence and relating it to neurology is also a rather tricky field. First you have to clearly state what intelligence is—which is itself no mean feat. Is it thinking faster? Knowing more facts? Is the doctor who passed medical school yet has never been abroad smarter than the traveler who can't do calculus yet can speak ten languages? They certainly possess different cognitive skills, but we'd need to agree which of these constitutes intelligence before we can determine a way to rank them.

Ever since people first started to study psychology, tools that provide a measuring stick for intelligence have been constructed. "Intelligence Quotient" (IQ) tests come in many forms, most of which have been modified over time to account for cultural differences in an attempt to quantify a single brain's biological talents. These tests tend to compare you with everybody else in your population—if you have an IQ of between 90 and 100, you're pretty average. An IQ of 120 means you're smarter than the average bear; over 140 makes you a genius.

The British psychologist Charles Spearman suggested early in the 20th century that the fact a school student who does well in one subject tends to do well in many might be caused by a dominant psychological trait called the "general intelligence factor." He used statistics to find common ground between different intelligence tests and developed a model to explain why they often came up with different results.

Psychometry (measuring intelligence) is a rather controversial field. In the past, such testing has been used to justify many forms of prejudice. There is evidence suggesting that aspects of intelligence can be inherited, or related to your family's background. Questions remain on how much of your intelligence is genetic and how much is caused by the environment you're raised in. Some researchers even claim that measuring intelligence is intrinsically impossible.

An issue that has long plagued research into intelligence is the association of superiority with vague notions of advanced evolution. As Charles Darwin's explanation of how evolution might work gained popularity in the latter half of the 19th century, anthropologists began to question whether some races were "more evolved" than others. Even Darwin himself speculated whether populations such as Australian Aborigines could be more closely related to apes than white-skinned Europeans. Some, such as the 19th-century American social theorist Lewis Morgan, linked social progress with innovation, associating acts of innovation with advancement out of savagery into civility.

Given the recursive nature of using a process of thinking to investigate how we think, it's obvious that studying intelligence isn't like studying any other biological process. Our difficulty in defining and measuring intelligence in an objective fashion has interfered with our ability to see it as an evolved feature little different from our lack of body hair or burgeoning ability to digest lactose. Our brains haven't just set us apart from other organisms—they've allowed us to climb an ivory tower that distances us from them, where some people are seen as having climbed higher than others.

Intelligence has long been linked with the evolutionary progress of primates, as if it's a marathon and humans are clearly in front simply because we have the ability to describe our cousins, the ape and the monkey. Historically, we've used technological progress as a marker for evolution within our own species, viewing cultures as primitive—and even less evolved—on grounds of correlating a people's innovation with their mental aptitude.

Was Einstein's brain more highly evolved than ours? Does it even make sense to ask such a question? More importantly, did humans evolve to be capable of deducing the secrets of the universe, or is it simply a fortunate accident—a side effect of an organ that adapted to suit a very different purpose?

FROM BUG BRAINS TO BIG BRAINS

According to Aesop, ants make a good metaphor for wisdom and foresight. His fable of the ant and the grasshopper tells of the benefits of hard work in anticipation of times of hardship. Then again, Aesop probably never saw what is called an "ant mill" or, in more common parlance, an ant death spiral. First described by the animal behaviorist Theodore Christian Schneirla in 1944, this fascinating phenomenon occurs when a trail of ants lose their way and double back onto their own pheromone trail, where they proceed to follow it again in a loop. The ants are programmed to continue to lay down their scent for their brethren to follow, further reinforcing the colony's desire to wander in a circle until eventually those trapped by it expire out of sheer exhaustion.

It's impossible not to sympathize with the poor little creatures dropping one by one on their road to nowhere. After all, we have a tendency to personalize the efforts of other organisms. To think of them slogging it out each day, working hard in search of food, it's easy to imbue the ant with a conscious will to act.

Ants aren't capable of applying reason to their actions. No ant is going to pause and ask their neighbor, "Didn't we pass that same twig not five minutes ago?" or look ahead to conclude that the end is nigh. Their inevitable starvation is the result of simple neurological programming that tells the ant to go where the smell of another ant is strongest, as it will probably lead to food.

While there is a world of difference between an ant and a type of blood cell in our body called a leukocyte (or white blood cell), they both behave as a result of competing chemistry. When a virus causes its host cell to rupture and spill its contents, a gradual diffusion of cell innards through the body's tissues creates a trail of breadcrumbs for the body's immune system to follow. Just as the nervous impulses of an ant occur under the direction of an external chemical scent, white blood cells move according to a behavior called "chemotaxis," where chemical signals set off a domino effect of reactions that result in a blood cell changing its shape in such a way that it moves toward the site of damage.

While chemistry explains both the nervous system of the ant and the bulging pseudopod of the white blood cell, each system has evolved for different purposes.

Single-celled organisms can usually deal with a stimulus by relying on simple changes to their internal chemistry. However, for much bigger organisms, the chemical reactions need to be coordinated with other tissues in the body, which involves communicating with different types of cells. Chemicals called hormones can carry this signal far and wide through the body's vascular system. As long as there's no big rush, pumping hormones through the blood works nicely. Plants do just fine using hormones to control how they feed, defend, grow and reproduce. We also employ a range of hormones to control our growth and development.

Yet more precarious situations might call for a more speedy reaction than a steady flow of hormones can provide. For example, moving away from a predator or toward prey could be a matter of extreme urgency requiring the coordination of a number of different organs. This often demands more than a simple "on or off" reaction—for instance, the organism must find prey if it is night-time, while hiding away from predators if it is daytime. These situations require far more speed and computation than the slow diffusion of chemical signals can manage on its own.

Neurons are cells that work together to create complicated behaviors. Their chemistry is capable of taking into account different stimuli in production of a rather immediate result. Whether it's an ant negotiating a pheromone trail on the way to a dead bug or a chimpanzee pulling termites from their nest using a twig, both of their nervous systems serve the same purpose. They rapidly deal with a complex situation to elicit a swift reaction.

Fundamentally speaking, our talented brains continue serve the same basic function as they have since those first nerves twitched all those millions of years ago—they help us survive in a constantly changing environment full of other organisms also doing their best to survive. They do so through coordinated chemistry, relying on a competition of signals to produce a reaction. The brain of the ant might not be capable of jumping the rails and ignoring the drive to play follow-the-leader to avoid a slow death, but in many ways our own nervous system is equally incapable of ignoring its underlying programming.

THE LAST OF OUR KIND

We *Homo sapiens* are the last of the hominins.

Australopithecus anamensis, Australopithecus afarensis, Paranthropus robustus, Australopithecus aethiopicus, Homo erectus, Homo habilis, Homo georgicus, Homo heidelbergensis, Homo floresiensis. . . . Rather than a single chain of human ancestors, hominins belong to a branching family tree—of distantly related cousins, uncles, aunts and grandparents—that fragmented and meandered its way out of Africa and across the globe.

Now, only a single twig of this tree remains. Ours.

Surprisingly, *Homo sapiens* didn't just survive where other hominins died off—over the millennia we've flourished to colonize nearly every environment imaginable on Earth, and have shown we can even survive in the hostile vacuum away from it.

At some point roughly six to seven million years ago, hominins closely related to modern humans were experiencing a gradual increase in brain capacity from about 350 cubic centimeters (equivalent to a modern newborn) to today's adult average volume of about 1,200 cubic centimeters. While an increase in the size of our skull's internal cavity does not automatically mean an increase in intelligence, there were almost certainly a number of neurological changes that became possible with an expanding brain mass. The bottom line is that as our brain size increased in volume, humans developed a far greater range of mental tricks than any other organism before or since.

It is rather arrogant to suggest any one of these traits, on its own, can't be found elsewhere in the animal kingdom, however. Language, depending on one's definition, seems to have rudimentary equivalents in many species. Crows and chimpanzees are quite competent at making and using tools. There is even some evidence of certain species of scrub jay being able to anticipate future events, giving them a rudimentary concept of "time."

Yet we certainly outperform other organisms when it comes to the complexity and depth of many of these mental talents. Other animals might use particular sounds to communicate, but humans seem to be the only ones capable of syntax and grammar. Some species of monkey might well bark a sound that warns "snake!" but without a system of interchangeable codes and variable contexts, it is more comparable to a human laugh or cry than it is to our poetry and stories.

A crow can cut a leaf into a shape useful for pulling a grub out of a knothole, yet humans are the only species capable of incorporating a range of tools with different uses to invent a single new tool for a novel problem. Even if the scrub jay can plan meals for tomorrow's needs, we can extend this "future sight" to anticipate the plight of generations not even born and dream of the sort of world they might live in.

The more we learn about the intelligence of animals, the more apparent it becomes that there is no single mental trait we can call uniquely human. Instead, we have merely evolved a number of those same features to a far more complex level.

We can be quick to glorify ourselves as the perfect animal, a pinnacle of evolution, holding up examples of our amazing technology and swollen population as proof. Yet evolution has no goal. It is a blind process, chugging along under the influence of a feedback loop between random variation and environmental change. If an arrangement of genes works in a particular environment, it will persist. As far as nature is concerned, we're no more evolved—or superior—than crows, scrub jays, chimpanzees, E.coli, crocodiles or shiitake mushrooms. Simply staying alive is good enough as far as Mother Nature is concerned.

The skills that afford you the ability to read this book and contemplate the meaning of its words are accidents of chemistry. Somewhere in your biological past, your ancestors were not only born with nervous systems wired a little differently from their siblings'—they survived and produced more offspring as a result. We now use these nervous systems to dream about traveler to the stars, to negotiate pay rises with our bosses, to give advice to our best friends about break-ups, to fix up a tasty lamb ravioli for dinner, to design sports cars, to weave plots set in fantasy worlds and to anticipate changes in the economy when buying a new house.

The complexity of our brains emerged only because it allowed our ancestors to survive and make more babies. As such, our brain is fundamentally no more special than our digestive system, or our heart or our kidneys—it behaves as it does simply because it has proved useful within its environment.

In the middle of the 1970s, a certain fossil dug up in the Awash Valley in Ethiopia was making all the headlines. Titled "AL-2881," she was introduced to the world as "Lucy." This assortment of bones was hailed as everything from the "missing link" between man and ape to "the first human." Neither of these speculations meant a great deal to the scientists studying human evolution. Each new-found fossil can be thought of as a missing link between its parents and its offspring, after all. However, the skeleton did have a brain size closer to that of modern apes, while it had the upright stance of modern humans. This blend of characteristics helped pinpoint where in our ancestry our hominin traits might have arisen.

Lucy is old news. Others have joined her esteemed ranks of fossil informants. *Ardipithecus ramidus*—Ardi for short—is where it's at today.

Ardi's fossilized remains were unearthed as far back as the early 1990s, but it wasn't until 2009 that details of the bones' analysis were published. According to his researchers, it's unlikely that you could ever trace your genealogy directly back to his family picnic; however, similarities between your ancient relatives and his physiology means he can be placed somewhere among the more remote branches of your family tree, living about 4.4 million years ago. Since Lucy walked the African plains only 3.2 million years ago, Ardi is one step closer to that important fork in the tree where chimpanzees and humans went their separate ways.

Being an older specimen, Ardi's remains reveal a lot more about those common ancestors. Ever since the arthritic remains of an old Neanderthal were uncovered in France in the early 20th century, there has been the misconception that primitive man must have walked with a stoop and looked like a cross between a man and a monkey. Following this incorrect assumption, the further back in time we go, therefore, the more "ape-like" we must have looked.

Ardi's skeleton doesn't look anything like a chimpanzee's, which is odd considering the human and chimpanzee lines diverged sometime in the couple of million years before *Ardipithecus ramidus* walked the Earth. Notably chimp-like qualities, such as evidence of walking on their knuckles and large male canine teeth, were absent. Ardi walked straight on two legs and had smaller, more generalized teeth for eating a variety of foods. His most ape-like trait would be his feet, which were adapted to grip branches.

What does this tell us? For one thing, it says that over millions of years, the ancestors of chimpanzees evolved to look like chimps while the ancestors of humans evolved to look like people. Chimpanzees don't look like ancient humans and ancient humans aren't simply human-like chimpanzees. We both evolved to become what we are today in response to the pressures of our respective environments.

Likewise, chimpanzee intelligence evolved to suit the chimpanzee's environment, and human intelligence evolved to suit a different set of circumstances. So why is it we can use our brains to play Pokemon and race sports cars while our chimp cousins . . . well . . . can't?

THE GREEDY ORGAN

If you're sitting back right now, relaxing, a bag of salt-and-vinegar chips in hand, 25 percent of the glucose in that tasty snack will be devoted to your gray matter. Out of every breath you suck in, 20 percent of the oxygen will contribute to keeping your thoughts ticking over. And yet, if you're an average male with a mass of 75 kg, your brain accounts for a mere 2 percent of your total body weight.

The brain is one greedy chunk of meat. If it were a housemate, you'd hope it provided a good share of the rent. Fortunately for many of us, food is plentiful and we needn't exert ourselves physically for our next meal. So your brain's ravenous demand for energy mightn't be of much concern.

The modern luxury of having a relative wealth of resources at our disposal hasn't always been the case. In a dog-eat-dog world, the kilo-joules contained in every meal we consume would be weighed against those expended in chasing it down or digging it up. There is no room for waste—every joule of glucose must contribute to the competition for resources, making babies or simply staying alive. While enjoyable, writing an opera or contemplating the nature of dark matter hardly accomplishes any of these survival functions.

This presents a distinct problem. To sustain the growth of a power-hungry organ such as the brain, early humans needed energy to devote to it. Where did it come from? While there is no single clear answer, there is no shortage of hypotheses on the topic.

One possibility is that a group of human ancestors stumbled across a new environment containing an abundance of energy that required little effort to harvest. Stephen Cunnane, author of the book *Survival of the Fattest*, suggests this new diet might well have been found on the ocean shore in the form of seafood, which would have been relatively easy to find and high in the right fats and proteins.

Perhaps another evolutionary change, such as the energy saved by walking upright, freed up just a few extra joules to allow for a tiny increase in brain mass, which in turn opened the way to even more energy sources. Similarly, perhaps a more efficient means of organizing our brain started a positive feedback loop, where a better connected (but not necessarily more hungry) brain initially allowed for improvements to our cognitive "software," leading to further savings in energy, allowing for more growth, and so on in a cascade of brain improvements.

This presumes, however, that a bigger brain automatically leads to a handy advantage simply on account of being bigger. In other words, it suggests that it's a law of nature that more is always better, and if you're lucky enough to have plenty of food or extra energy, there's a good chance you'll devote it to neurological development. This is called the "epiphenomenal" hypothesis of brain evolution, and has been suggested by scientists like Cornell University psychologists Barbara Finlay and Richard Darlington. They compared the brains of numerous mammals and found that aside from the tissue associated with the sense of smell, there is a general matched increase in the brain's entire size regardless of which specific behavior is helping the organism survive. For example, better vision doesn't just mean more nervous tissue to see with—it means other parts of the brain necessarily expand as well.

To better compare the brain sizes of different organisms, a figure called the encephalization quotient (EQ) is used. This ratio helps contrast a mouse's brain with a whale's brain in relative terms. After all, as whales have more body mass, you'd only expect their brains should be larger. So, by comparing an EQ rather than just comparing brains on a set of scales, it's like shrinking or expanding all animals to the same size and then matching big brains to big smarts. Humans and dolphins tend to rank quite high. Animals like the armadillo and kangaroo don't.

Using EQ as a comparison, it's possible to compare diets of animals and see if there is a relationship between the amount of energy they get and their brain size. Sure enough, primates that live on fruit do tend to have a larger EQ than those that live mostly off leaves.

However, just having access to plenty of energy is unlikely to be sufficient to automatically produce more neurons. There has to be more to the story given that evolution simply doesn't favor dead weight. If expensive brains don't actively contribute to the process of survival, it's unlikely they'd remain in competition with cheaper, simpler brains. Any "spare" energy would simply go toward making more offspring, meaning the plains would soon be swarming with efficiently stupid human ancestors over less efficient smart ones. Therefore, while brains do get bigger as an animal's body gets larger, and having an available source of energy is indeed a necessity, these facts aren't quite enough to solve our problem. That extra brain mass has to do something to either save the animal energy, or find more energy than it costs to run.

If all of the components of human intelligence contributed to identifying where the best berries would be found, how to not be eaten by angry lions, how to tell which flowers were poisonous or how to dig for water in a drought, it might pay off.

Unfortunately, the evidence fails to address why humans couldn't make do with simpler brains, especially if other species with similar niches do just as well without such extraordinary (and expensive) talents. In other words, if a chimpanzee had our mental acuity for precise language and progressive tool making, it's unlikely they would find significantly more food or avoid more danger as a direct result of those cognitive skills. At least, not enough to pay for all of that extra mental work.

Another rather interesting theory states that our gift of wit developed to woo our desired sexual partners. Before you laugh too hard, there are a number of traits across the animal kingdom that developed solely as a way of winning the heart (or other body part) of a loved one. Bright plumage, long hair and mating calls often developed as a way of telling potential mates that you were fitter and healthier, and therefore genetically stronger, than everybody else. According to the Hungarian zoologist Lajos Rózsa, being clever could be a way of showing the other sex that you're free of brain parasites and therefore have a superior immune system. Our brain could be the neurological equivalent of a peacock's tail.

There is one last possibility, however, that has a lot going for it. Intelligence, it appears, might not be about finding berries, fighting lions or demonstrating good genes, even if they all might contribute. Rather, the evolution of human smarts might be more about exchanging gossip, telling stories that dichotomize good and evil and making us feel sorry for having laughed at our grandfather when he stepped in mammoth poo.

Growing smarter brains might have more to do with gaining a benefit from manipulating and cooperating with other people in our tribe than dealing directly with the savagery of Mother Nature.

SPEAKING OF GOSSIP . . .

The study of glottology, or the origins of language, involves a mix of disciplines such as linguistics, ethnology, neurology and paleontology. For we humans, it involves a rather interesting "chicken or egg" dilemma— which came first, bigger brains or our ability to use words?

It's an important question. Language plays a vital role in our ability to describe the world in our heads. In fact, it could be argued that so much of our knack for comprehending our environment stems from our skill in mirroring the world with mental concepts using symbolic thinking. In this light, it's easier to relate our intelligence with our ability to turn ideas into simpler representations.

Fossil records suggest our ancestors could make a wider range of sounds with changes in the shape of their trachea, an event that occurred with the joys of standing fully upright. In other words, altering the position of our spine and hips repositioned our head and neck. This allowed our voice box to move lower, providing more room for a greater articulation of sound. This also provides a nice place for food to get stuck, increasing our risk of choking. The cost of gossiping with friends must be worth it! Such changes didn't occur overnight, but rather progressed slowly over several million years, hinting that the more sounds our ancestors could make, the more of an advantage they might have over less verbal members of the species.

Language isn't just a matter of making sounds that mean something, however. Many animals can squeak, squawk, croak and roar in communication, telling those in its own or other species that it is angry, hungry, horny or frightened. Rather, language involves a set of rules that govern the meaning of sounds as they are strung together in a variety of ways. It also involves the manipulation of symbols by making a mental model of the surroundings and then selecting commonly understood sounds to represent and communicate that model.

This is quite a big task for a nervous system to manage. Language relies on a lot of complicated mental computing, incorporating numerous neurological regions. Rather than looking at the brain as a single lump, it's important to judge how different areas might have developed in relation to one another over millions of years. Significantly, human brains aren't just bigger versions of those that occupied the skulls of their ancestors. Their change in shape over time reflects an expanse of some brain areas while others grew little, stayed the same or even shrank.

Good communication, both verbal and non-verbal, is useful when living in a large community of individuals who must cooperate in order to survive. In nature, there is safety in numbers. A group of animals can work together to find food and protect their young. If you can work well as a member of a family, you'll have a better chance of getting more food for less energy than the divided mob over the hill. Effective leadership can encourage a community to stand against an enemy, negotiate politics with neighbors or push into new territories when food is scarce. Within a community, good communication is the key to winning friends and influencing people.

Most social mammals negotiate intimate relationships by grooming one another. This also establishes a hierarchy, where those higher up the pecking order have burrs and parasites removed from their hair by more submissive members of their group. Not only does grooming reduce your flea infestation, it releases feel-good hormones such as oxytocin—think of how a hug, or even a gentle touch on the arm, makes you feel. However, it imposes a limit on how big your group can grow. There's only so much grooming anybody can do in a single afternoon, and some monkeys can spend up to a fifth of their day picking at another monkey's hide. Big groups have certain competitive advantages over smaller groups, especially in open environments, but you can't possibly be friends with everybody.

Psychologist Robin Dunbar suggests that language provides a more efficient means of exchanging social information in an effort to bond, particularly as grooming becomes more difficult to manage in increasingly larger social groups. Even today, we devote a lot of our time and effort in conversation to the exchange of intimate details rather than discussing survival techniques, with Dunbar pointing out in his book *Grooming, Gossip and the Evolution of Language* that 60 to 70 percent of conversations focus on relationships and personal experiences.

As our ancestors' social groups found advantages to congregating in larger communities on the open savannah, those who could vary their communication to replace the need for so much grooming rose up the chain of command faster, which in turn led to producing more babies. Put simply, as a species we have found significant advantages in working together in relatively large teams. Our ability to understand each other's account of what we've experienced allows us to get the most out of our social group, forming the means to pay for that expensive brain.

A rather simple example of the difference in how we humans communicate compared with our closest relatives is the humble gesture. Even toddlers can point their hand to help somebody find an object. Chimpanzees, gorillas, bonobos and orangutans have a range of gestures as part of their non-verbal vocabulary, yet pointing doesn't seem to be among them. At least, not naturally. They will clap their hands, thrust their hips and poke other individuals or pull their hair, but in the wild there are very few examples of one ape indicating where another ape should look.

That's not to say the behavior is beyond their abilities. Researchers at the Max Planck Institute in Leipzig, Germany, and more recently at the Great Ape Trust in the United States, have recorded the behavior of chimps and bonobos pointing for their human researchers.

The Max Planck biologists studied this curious difference between their behavior in the wild and how they act in human company by setting up a pair of barrels in a room and placing food in one of them. On letting a chimp enter the room, one of the researchers would point at the barrel with food in it. Over a number of runs, the experiment showed the chimpanzee selecting a random barrel to find the food. The pointing was irrelevant—the ape might as well have flipped a coin first, for all the good it did.

Something interesting happened when the researcher changed his gesture, however. By appearing to reach for the barrel rather than just point, the chimp seemed to understand. Although quite subtle, the difference was an action of competition rather than cooperation. Chimpanzee social behavior is not as cooperative as that of humans; there is less of a sharing of intent, of pointing just to assist another's perception.

We typically count intelligence as a trait that defines individuals. Albert Einstein was regarded by many as a genius, for example, while history is littered with examples of people who could be regarded as idiots, fools and generally—for lack of a better word—stupid. But we rarely pause to consider how groups of people work together to form an intelligent unit.

In 2010, a collaboration of American researchers from Carnegie Mellon University, New York's Union College and the Massachusetts Institute of Technology investigated how groups of individuals might cooperate to solve problems. They set out to test how a team might fare on a diverse range of cognitive tasks, and whether—as with a single person—there was significant consistency from one task to the next. On our own, if we tend to do well solving one type of problem, we tend to do well on a number of different mentally challenging activities. Working together in groups seems to be no different.

However, far from simply being about the weight of intelligence within a team, the researchers discovered problem-solving varied with social dynamics. Social sensitivity was a better marker of how well a team would do on a task than an averaging of the group's cognitive abilities. What was more, those groups dominated by one or two extroverted figures were revealed to be poorer at tasks than those demonstrating more efficient communication tactics.

The skills we as individuals rely on to solve problems appear to be different from those that we use as members of a community. An ability to empathize, identify emotional cues, communicate effectively and cooperate tends to serve us better when it comes to collectively solving problems than being lone masterminds.

The complexities of our speech are linked with our ability to cooperate socially. Our single greatest asset has less to do with how we as individuals deal directly with the wider environment, but how we use our brains to work as a tribe in collecting information and seeking useful patterns. In a rather broad sense, our brains have evolved as little more than storytelling machines used to share experiences that involve our friends, enemies, ancestors and, sometimes, even individuals who aren't other people.

MY PET MONKEY

When I was four, I had a pet monkey. He didn't eat much, but I walked with him everywhere I went (except to kindergarten, as I assumed he wouldn't be allowed inside). The fact that only I could see him didn't seem to be a problem. In my mind he was real enough, to the point that a parent of one of my playmates felt obliged to ask my mother if there was any truth to my story.

"Animism" is the term used to refer to the supernatural philosophy ascribing self-awareness or a will to inanimate objects and natural events. A simpler definition is the belief that most things—regardless of whether they are alive or not—have a spirit or soul. Animism has formed the basis for belief systems at least as far back as the beginning of recorded history, and probably those much older. Even today, many cultures and individual people are animistic.

This tendency to interpret the world in such a way is so ingrained it even represents a significant stage of childhood development. Swiss child psychologist Jean Piaget described the progress of psychological development in children using four key stages, each broken into several smaller periods. Occurring between the ages of two and four, the "pre-conceptual period" is defined by egocentric behaviors (where the child has yet to develop a way of understanding the world from another person's perspective) and animistic thinking. Imaginary friends and sympathy with objects such as teddy bears or a favorite blanket are common during this brief phase in a person's life as their brain adapts to working with other people, establishing the so-called theory of mind.

Simply put, this is the comprehension that you have a mind and so do others. A term used to describe theory of mind is "intentionality."

In linguistics, intention refers to the properties we associate with a word or a symbol. The Swiss linguist Ferdinand de Saussure describes words being made of three parts—the symbol itself, like a spoken word or a physical image, called the signifier; the concept that this word creates in another person's mind, called the signified, or "intention"; and the real-life thing that is being described, or the referent. Intentionality is the ability to grasp the fact that a thought or image can exist not only in your mind, but in another person's mind from a different perspective.

Computers have "zero-order intentionality." No matter how much information you program into their memory, there's no evidence that a software program is aware of its own information load.

First-order intentionality describes anything that comprehends its own state of mind, but doesn't understand that other things are also self-aware. This describes small children, who grasp their own knowledge but fail to comprehend it existing in another person's mind as well.

Minds with second-order intentionality know that others also have a mind, but stop short of understanding that those others, too, understand that others have a mind.

Third-order intentionality is tricky—that is, a mind that knows that another mind knows that they have a mind. In effect, we can keep going to fourth and fifth orders, where it gets hard to retain a good grasp of intentions. For example, I know that you know that your mother thinks that her sister believes that her friend hopes that her boyfriend considers . . . and so on.

The development of multiple-level intentionality in humans has played a vital role in the development of science. It enables us to comprehend the possibility that somebody else can see and think about what we see and think, and that they might also see and think something we cannot.

Our complex brains might have evolved to establish strong social bonds, but defining what should be part of that social group appears to be rather flexible. For instance, some people treat their pets as if they are members of their family, and believe they can relate to the pets' needs, desires, fears and dreams. Others see animals as distinct from humans and wouldn't think of talking to one as if it was a friend. Many people keep the ashes of a deceased loved one nearby, believing those ashes continue to serve as some link to their existence. Even if inanimate objects aren't imbued with a distinct personality, many of us still attribute certain human characteristics to non-human objects or entities, such as pets, dolls or even cars and memorabilia.

Have you ever taken something to be signed by your favorite musician, sporting hero or actor? A clipping of Elvis Presley's hair that was cut when he enlisted in the US army sold for US$15,000 in a Chicago auction house in 2009. For all appearances—microscopic and macroscopic—the hair looks no different from that you'd find on millions of other human heads around the world. Yet the person who bought it believed there was more to the hair than just keratin and a few flecks of dandruff. The fact it was once on Elvis's head gave it properties that no other hair has, connecting its buyer to the king of rock and roll.

This trait can go the other way as well. We're less likely to buy houses that have been the scene of a murder or to wear clothing that was worn by a person considered to be evil. Of course, this could be explained by an instinctual desire to avoid possible contaminants that caused the bad things to occur in the first place. We might have a better understanding of pathogens today, but having an innate drive to avoid potentially unhealthy situations would still be of great advantage.

Either way, attributing a personal essence to non-human things—from anthropomorphizing animals to believing that trees are self-aware or to thinking that wearing Adolf Hitler's hat could transfer an evil power onto you—is a feature of a brain that evolved to see a world in terms of personalities rather than physics, forces and chemical reactions.

It never truly mattered that my pet monkey wasn't a fleshand-blood animal. Its purpose was never to take the place of a real pet—the imaginary friend was the result of a stage in my development where I exercised my abilities to socialize and understand different perspectives. According to University of Oregon psychologist Marjorie Taylor, imaginary friends don't represent abrupt phases of development, but rather exist on a continuum of social learning. Their presence indicates social cooperation and scale of interaction with others.

Cultures that believe a stream is crying for a lost child, or that the wind is angry because of a missed sacrifice, aren't expressing critical attempts to understand the universe by means of geology and physical interactions. Rather, our brain's aptitude for symbolic thought, intentionality and empathy comes to the fore in our drive to find relationships between our observations. Somewhere on the same continuum as a child's imagined social group arises a tendency to view the world with a human face.

CHEATERS ALWAYS PROSPER

Jesus appeared to Mary and reminded her that "life is going to be good." No, not *that* Mary. This Mary lives in Massachusetts. But it was *that* Jesus . . . at least, it was a rust stain on the bottom of an iron that partially resembled a face framed in long hair.

If Mary wasn't Catholic, the rust stain might have reminded her of some other long-haired personality, such as John Lennon. If you squint, the brownish smear bears a faint resemblance to the Mona Lisa. But Mary's faith meant there was only one conclusion she could draw from the strange image she noticed on the bottom of her home appliance.

Why would a face appear on the gray metal panel of an electric iron? An oxidative reaction between the ferrous particles in the appliance's alloy plate and the atmosphere, most likely. Years of combined physical wear and electrochemistry caused a crescent-shaped, brown-colored pattern to form. But why a face?

The answer lies deep inside your brain, in a section called the *fusiform gyrus*. This tiny patch of nerve tissue has the task of identifying faces for you. Given a pattern of shades, lines and colors, it will find the familiar shadowing of the eye sockets, identify the ratio of creases that forms the mouth or nose and latch onto dark smears that could identify a hairline or cheek. Thanks to this patch of tissue, faces can quickly be recognized as familiar or foreign. In those who lack such a tribal-recognition system—a condition called prosopagnosia—faces aren't so easily recognized, making strangers of friends and family.

Mary's iron might have produced any one of an endless variety of different shapes. Some might have looked like a tree, or a hamburger. Most wouldn't have looked like anything at all—just vague blotches, speckles and streaks. In fact, it's more than likely that Mary's house is full of similarly random patterns. Yet, by chance, some will resemble a shape we can identify. Our pattern-matching brain does the rest—a tile in her kitchen might have a burn mark shaped like the letter "P"; a stain on a towel might look like a teddy bear. She could have a banana with a black bruise that resembles a submarine.

It's a good thing the brain can find these patterns. Tigers, bears and crocodiles don't come with name tags and flashing lights, so you need to be able to identify one faster than it can catch you. Most shapes in the world that look a lot like tigers are indeed great big, stripy killer cats. However, sometimes a messy shadow in the undergrowth will coincidentally look like a tiger. This is the brain making what is called a "type 1 error," or identifying a false positive. This occurs when a pattern, such as dark stripes, is mistakenly attributed with meaning, such as causing you to believe it is a tiger. "Type 2" errors are the opposite, when a meaningful pattern is dismissed as useless. Nobody ever died by running away from an innocent shadow, but there is plenty to lose if you mistake a tiger for a bunch of reeds.

However, in this instance, Mary saw a face that didn't mean anything important. At least, no more than a cloud that looks a bit like a carrot. Unlike teddy bears, submarines and the letter "P," an essential shortcut in Mary's brain added significance that screamed "faces mean people." In nearly every other instance, it would be correct. It would successfully identify the face of a celebrity in a grainy black-and-white photograph, a caricature of the prime minister in a cartoon or Mary's husband as he approached from a distance.

Influenced by her religious beliefs in the supernatural, Mary felt an overwhelming urge to attribute a personality to what was otherwise a common stain, rather than to just smile and show it to her husband. And not just any personality, but one of immense importance. For her, Jesus was the most important personality she could think of.

This is Mary's brain taking a gamble, as all brains should. If they didn't, they would need to be a lot bigger, a lot hungrier and probably a lot slower. They would need more information to act and a lot more processing power to deal with that mass of information.

Since most of the face-like patterns Mary sees every day are indeed faces, her brain's gamble usually pays off. In addition, making the mistake of thinking there is the face of Jesus on her iron isn't costing her much (except a new iron. Nobody likes to do housework with the face of God).

Your brain deals with a lot of information constantly streaming in through your senses. To make sense of every last scrap of it would be impossible. So it plays a numbers game and gambles on the outcome. Shortcuts are used to make assumptions that are usually correct. But not always. We frequently make type 1 and type 2 errors. The brain might not be correct 100 percent of the time, but given a choice between being wrong on rare occasions and needing a head the size of a whale (not to mention a diet to match), which would you prefer?

Far from being perfect record-keeping devices like some magnificent computer with a huge hard drive, our brains are cheating, lying bits of tissue that serve mostly to cooperate with our friends and family in an effort to get the most out of our environment. Our brains are tribal engines, evolved to think in terms of faces and feelings rather than facts and physics.

Unfortunately, if you think that means you can relax and not care about whether it is accurate or not, so long as it helps you to survive, think again. You couldn't be so lucky.

TURN LEFT FOR ANSWERS

On looking at a human brain, one of the first things you'll notice is that there is a big split right down its middle. From above it looks like a pair of giant, white, wrinkled prunes squished together. The "prunes" are called hemispheres, and you may have heard that each hemisphere is responsible for controlling the opposite side of the body; the left half of the brain takes care of the right half of your body, and vice versa.

Scores of books have been written on the "left brain–right brain" phenomenon, insinuating hidden creative talents which the reader can take advantage of if only they can tap those neurological reserves. While it mightn't be entirely incorrect to describe the left hemisphere as "logical" and the right as "creative," it does dramatically oversimplify the situation.

The truth isn't quite so neat. Each half of the brain does control a number of muscles and functions on its opposing side of the body. If you have a stroke in the left hemisphere, you could lose the ability to move your right arm or leg, or even the muscles on the right half of your face. For the most part, images you see with the left eye are sent to the visual area of the right hemisphere, and images seen with the right eye are sent to the left. However, some involuntary muscle movements in the face are controlled by both hemispheres. And while the left and right hemispheres do respond to events in their own ways, neither can be summed up in simple words like "reasonable" or "emotional." Creativity—in all its complexity—relies on a variety of neurological traits found all over the brain.

Michael Gazzaniga is the director of the Student Achievement Guided by Experience Center for the Study of the Mind at the University of California. His work has involved studying the behavior of individuals who have undergone a radical neurological procedure.

Stretching across the thin divide between our hemispheres are filaments called the corpus callosum, which allow information to be exchanged. For some epilepsy sufferers, seizures are so severe they can ripple across this bridge and place the sufferer's life at risk. In cases where pharmaceuticals are inadequate to control seizures, cutting the corpus callosum effectively separates the hemispheres, preventing the seizure from spreading and reducing its severity.

So-called split-brain patients have provided researchers with a prime opportunity to evaluate how each hemisphere deals with the world, and what happens when one side literally cannot communicate with the other. For all purposes, these people have two brains that operate in complete independence without any awareness of their partner.

You'd be forgiven for wondering if such people had split personalities. For instance, is each half of the brain aware of itself? Can you read two books at once? Are there two patients in the one skull?

While awareness is something that arises between the two halves (your own thoughts aren't specifically located on one side or the other, but rather is the sum of a range of functions spread across a number of different regions on both sides), splitting the brain won't dissect your consciousness at all. The halves still receive information through the brain stem and spinal cord, and are each aware of the same world outside of the individual's body. Both hemispheres keep doing what they've always done. They just can't share the results of their work.

For the most part, the split-brain patient operates no differently than any other person. They might unconsciously move their head a little more in order to provide both hemispheres with more information about their surroundings, or speak in a loud voice to compensate for the reduced communication in their heads, but for most purposes their brains carry on life as usual.

However, under certain circumstances, the hemispheres can reveal some rather surprising secrets about how they each rely on the other to create a perception of their surroundings.

Professor Gazzaniga presented one of his split-brain subjects with two pictures separated by a division, allowing each eye to see only one image. For example, in one case a chicken claw was shown to the right eye and a snow scene was shown to the left eye.

The subject was then asked to pick a second pair of pictures from another selection that was completely visible to both eyes. However, they were required to pick one image with their left hand and one with their right. Not surprisingly the subject chose a snow shovel with the left hand, corresponding with the snow scene observed by the left eye, and a chicken with the right hand, matching the chicken claw seen by the right eye.

On being asked for an explanation for the selection the subject's response was remarkable. The chicken was obviously chosen as a match for the chicken claw . . . yet apparently the snow shovel was for cleaning out the chicken's coop.

Why the obsession with the chicken? What happened to the snow scene?

When asked to provide reasoning for the chosen pictures, the split-brain patient used their left hemisphere to come up with an answer. Because the right eye sent the information about the chicken to the left hemisphere, that was all it saw. Neurological tools within the left hemisphere are biased toward interpreting information within a meaningful context; the subject was relying on this half of their brain to answer the question. The left hemisphere didn't have a clue about the snow scene because its connection with the right hemisphere had been severed.

While the right hemisphere could still link a snow shovel to the snow scene, it wasn't able to provide a meaningful reason for the link it made. The left hemisphere had to make a guess as to why the other hand chose a snow shovel, based on the only information it had access to—a picture of the chicken.

The experiment was repeated with a variety of images and using a range of emotional states aimed at solely the right hemisphere, such as encouraging laughter or sadness. The end result was the same—the subject would experience a situation with the right hemisphere and be forced to find a reason with the left one, even if the left hemisphere had no information to go on.

Put simply, Gazzaniga's experiment showed that "I don't know" is simply not an answer the brain is happy with. We have evolved brains that go to great lengths to find a context for what we sense: we will fabricate stories based on our shared experiences and communicate them effectively with others, even if the information is rather limited.

Again, it is a case of the brain taking a gamble. Hesitating and waiting for more information before you act can cost precious time or resources. Sometimes, this gamble is a bad thing. Yet on the balance of numbers, it is very rewarding.

THE STORYTELLING MONKEY LIVES ANOTHER DAY

Humans have eyes that can only detect colors of wavelengths in the range of approximately 380 and 750 nanometers. We can't see the amazing ultraviolet patterns bees can see on many flowers, or the hazy infrared glow of a warm body. Our hearing is limited to vibrations little slower than 20 per second and rarely much faster than 20,000 per second. The call of a bat as it seeks its prey is silent to our ears, while the rumble of an elephant in the distance as it calls to others in its herd might only be felt as a subtle caressing in our guts.

It's difficult to truly appreciate the tiny dimensions of quarks within an atom's proton or enormous clusters of galaxies at the outer reaches of the universe. The slow grind of evolution over eons might as well be the relatively rapid pace of the continental crust moving across the mantle for all we notice. This limited range of perception has allowed our branch of the family tree to survive in a very narrow section of the universe, dealing with the sudden small but significant changes in environmental conditions we've faced across the millennia.

Our brains do well to help us survive on this tiny island of a planet. What is more, our cognitive tools weren't shaped by a need to describe our universe in perfect detail, to count neutrinos or see the radio waves emitted by distant stars, but rather by a need to negotiate our way through family squabbles and work within our tribal relationships to collectively deal with the unpredictable elements of nature. We're primed to quickly recognize a face, to pick words from a cacophony of sounds, to attribute emotions and thoughts to other objects. On occasions when these talents malfunction, whether through a sensory loss or a disorder such as autism, individuals find themselves alienated from the community, handicapped in their attempt to connect and engage with others often to the point of isolation.

A combination of factors in our social brains makes it possible to invent stories that go far beyond the personal nuances of fairy tales. Hungry for answers, we're driven to reconcile the things we observe with the beliefs we inherit. In recent centuries, the stories we tell that describe our universe have become more detailed, more permanent, recorded first in stone, then on paper and now in electrons. The beliefs themselves continue to evolve under the pressure of logic, passion, desire, hope and reason in a system we call "science."

Yet for most of human existence, our beliefs weren't formed under the influence of reason and consistency. The stories we told weren't open for criticism or comparison. Something changed that led to a new form of storytelling, with new rules and new values.

Those tribal forces that shaped our brains over countless millennia changed just enough about 2,500 years ago to create a spark that would revolutionize how humans saw the world. Our social landscape might have changed immensely since then, but our brains haven't changed much at all. We continue to see the world using a brain that evolved to tell tales of sad mountains and see faces in the passing clouds.

Chapter 2

THE CREATIVE SERPENT

Where did science come from?

S cience is a relatively new way of telling stories. Shaped by discussion of facts and observations, mediated by logic and reason, it produces stories that conflict with the myths passed down by our ancestors. While our stories are the media through which we frame a belief, science comprises the values against which we assess whether a belief is useful at explaining what we observe. Looking back into humanity's past, however, shows science to be a very recent way of thinking. For most of history, a very different set of social forces shaped the beliefs that founded the stories we told about our universe.

IN THE BEGINNING . . .

In 1926, the British anthropologist Alfred Radcliffe-Brown noticed diverse Australian Indigenous cultures each recounted tales of a giant serpent whose immense body carved mountains, rivers and lakes from the terrain as it slithered across the land. In the dialect of one particular Brisbane tribe the snake is called "Targan." In the north-east of Queensland, around Boulia, it is known as "Kanmare." Variations of this story can be found as far south as Victoria, but they all have one thing in common: the migration of a giant snake explains the formation of key features of the surrounding landscape.

Not all stories describe what the snake looked like. Radcliffe-Brown was initially interested in the myths told by families in the Arnhem Land area, and therefore called this snake the "rainbow serpent" after versions that depicted it with meteorological attributes such as a rainbow for a body, a thunderous voice and a tongue of forked lightning. Some stories associate the snake with the band of stars we call the Milky Way, therefore also linking it with the creation of the night sky. In many cases, the serpent is said to still exist, resting nearby in a sacred spot either as a threatening presence or with the promise of divine knowledge for those who seek it.

The travels of the rainbow serpent take place in the Dreaming; a realm common to a great many Aboriginal cultures, it's a time when the land and all it held was inhabited by kindred spiritual forms.

Depending on your cultural background, it can be somewhat difficult to appreciate what is meant by "a time" in this particular context. The Dreaming didn't occur during a particular period of history, as such. Nobody can count the years since it ended, and in some ways it is ongoing. This makes sense in cultures that have no means (or need) to accurately record a linear progression of events as they occur throughout the generations. Yet if you're used to thinking of today's date as a box on a calendar that stretches back to the beginning of the universe, this aspect of the Dreaming mightn't make a great deal of sense.

The question is: do people believe such stories are accurate accounts of how the world really is, or how it really used to be? Or are they merely allegories, providing us with a vague sense of creation in a metaphorical context?

The word "myth" is often taken to refer to fiction or to stories that are factually incorrect; however, it's not hard to find two people who disagree on whether the resurrection of Christ is a fictitious metaphor or a historical event. Therefore, one person's myth can be another's "factual" event, depending on how a person constructs their beliefs from their experience.

Ask an Indigenous Australian brought up on stories of the rainbow serpent whether a giant snake "really" made the mountains and rivers and you'd probably get a confused look. Sure, movements in Earth's mantle might have pushed the crust along a geological fault. Yes, the river is probably the result of water weathering the crust as it descends under the pull of gravity. So what? That doesn't mean the rainbow serpent doesn't exist . . . it just explains the land's creation in a different way. Did the Judeo-Christian god really create the universe in seven days? For some, he did . . . and the universe also formed by way of a rapid expansion we describe as the "Big Bang." Genesis simply describes the creation of the universe in a different way.

Scientifically, this simply can't be so. Metaphorical models strive for precision in their terms, progressing from being analogical to denotative, shedding unnecessary symbolic meanings where possible. A particular mountain, lake or beach was made either by a combination of geological activities or the movements of a giant snake. It isn't possible that both accounts can be equally true. So, too, the universe could not have taken both seven days and 14 billion years to develop. While mythological accounts don't have to be mutually exclusive, scientific stories do.

Science and mythology are two ways we can describe reality. Each describes our universe from fundamentally different sets of principles. Myths are narratives that attempt to connect our personal lives and those of our friends and family with our surroundings. They aren't accounts of a practical history of the universe; rather, they are stories of a world that seems to be a part of our tribe, perceiving everything in terms of good and evil, moral and immoral, friend and enemy. Myths are emotional, personal and quite often didactic, providing meaning through the decisions a protagonist makes as they encounter a dilemma.

Of course, myths do contain important pragmatic information about the world, occasionally serving as warnings or highlighting important annual cycles such as harvests or migrations. The context, however, is nearly always from a sentient being's perspective. Natural events are explained according to the choices made by some powerful entity and can therefore be understood in terms of emotions and social interactions. Those beings—be they omnipotent gods, local deities, malicious spirits or guardian ancestors—can typically be reasoned with on some level. Gifts can be bestowed or negotiations exchanged in an attempt to influence future events. Similarly, doing the wrong thing can incite rage and lead to punishments such as holding back important rain or releasing a devastating plague.

Early anthropologists were intrigued by the similarity between the myths of unrelated cultures. More than one creation story describes the early world as a featureless expanse of rock, or an endless, insipid ocean. A willful act then establishes order, either by destructive means, such as breaking an object or killing an enemy, or creating something out of nothing. Other similarities include catastrophic, regenerative events like floods that "cleanse" the world, allowing it start afresh; families or communities of gods that are overthrown by newer deities; and sacrifices that result in the creation of something new.

In a surprising number of cultures, the Pleiades constellation is described as a group of women. According to Greek mythology and one version of an Australian Aboriginal story, the stars are all sisters. In Nepal they are considered to be sisters-in-law and a lone brother-in-law. In Hindu they are mothers. It's strange how so many cultures interpret the same cluster of tiny pinpricks of nocturnal light as not just people, but as females. Perhaps this recurring theme is just a coincidence? After all, in other cultures, the Pleiades are represented as men, wheat or even chickens. Or is this trend indicative of something fundamental about how we create our stories?

It's possible that myths from diverse cultures share common features because ultimately we all share the same social relationships.

No matter what your background is, it's a sure bet that you have friends and relations you like and acquaintances and enemies you don't. There are top cats and underdogs, "haves" and "have-nots," heroes and scoundrels. You fall in love, have babies and mourn one another's death. It doesn't matter if you live in the desert, in the Arctic chill, in a massive city, on the edge of a vast savannah or on a tropical island—fundamentally, you'll have a great deal in common with every other person on the planet. Naturally, the stories we all tell would reflect these same social forces of love, death, anxiety and competition for resources.

The 5th-century BCE mythographer Euhemerus speculated that myths were typically based on actual historical events, even if they had progressively become metaphorically twisted in their retelling. To him, the Greek gods developed from tales told about celebrated people whose actions were retold with embellishment. Euhemerism relies on the presumption that the fundamental elements of a myth necessarily arise from an actual observation or experience, rather than using nonspecific events in a generalized fashion to inspire a story.

An ancient example of such myth-rationalization can be found in the

13th-century Norse fiction the *Prose Edda*, which describes the origins of the gods as warriors from a distant land. This "normalizing" of mythological accounts continues to be embraced today, with scholars attempting to connect science and mythology through specious possibilities of solar eclipses to explain sudden darkness, comets to explain celestial omens, algal blooms to explain transmutations of water into blood or diseases to explain demons, curses and magical possession.

Research conducted by the US National Center for Atmospheric Research considered whether the phenomenon of "wind setdown" was a possible hydrodynamic explanation for Moses crossing the Red Sea, as described in Exodus 14. In essence, the paper makes a reasoned argument supporting observations in a biblical story that is otherwise considered to have the hallmarks of a myth. Their aim echoes that of the 17th-century German astronomer Johannes Kepler, who calculated that a conjunction between Jupiter and Saturn could account for the appearance of the Star of Bethlehem over the legendary birth of Christ. Accounts of mythological beasts are regularly associated with real animals, such as unicorns with the toothed narwhal or dragons with monstrous monitor lizards.

While it's possible that many myths could well have started their existence as an impressive anecdote about a specific historical event, there is no reason to suspect this is a necessity, or even a common form, of myth-generation. One need only look to the process and purpose of modern storytelling, where even movies that claim to be based on true stories can as often as not be hybrids of diverse events and pre-existing formulas and clichés.

If myths serve a purpose other than to accurately record and describe how reality operates, it shouldn't matter if they conflict with one another or contain seemingly impossible feats. It's common in myths to show no regard for physical constraints or for contradictions between narratives. Events within the Christian gospels contain numerous contradictions, with all being equally true to many believers. In Greco-Roman legends, a god can emerge fully formed from a hole in another god's skull without the storyteller needing to account for differences in size or the fact that biology prevents people from being born that way. Animals can talk or transform, the dead can return and time can lose all sense of meaning.

This tolerance for inconsistency within myths, and between myth and the observed world, indicates myths aren't created with a single, consistent model of the universe in mind. Rather, they are the result of our social brains doing what they do best—telling personalized stories that are easy to relate to. Myths matter by way of how their characters feel, how they relate to one another, what struggles they face and what consequences they suffer. They strive not so much to describe the "how" of what we observe, but rather to ascribe reasons to "why" things are the way they are.

For most of our history, the universe was an intimate part of our community. A common feature of numerous pre-industrial cultures is the familiar relationship between people and their surroundings. What changed? What led certain cultures to invent a different way of thinking about reality, one that strove to explain nature objectively and was critical of the personal stories of old?

A little over 2,500 years ago, for perhaps the first time in history, the social environment in a tiny corner of the globe underwent a revolution. Far from a single event, this progressive change was a culmination of factors that coincided on the Mediterranean coast in what is modern-day Turkey.

THE SEEDS OF SCIENCE

If you were to visit the ruins of Miletus you'd find it hard to believe that it was once a bustling harbor city. Since its settlement during the Stone Age it has passed through the hands of a diverse array of cultures, only to be abandoned by the Ottoman Turks in recent centuries when its harbor filled with silt.

Today, the Aegean Sea lies ten kilometers to the west. Weathered stones of an ancient amphitheater, a few paved roads, a smattering of fallen columns and a handful of crumbling stone walls are what's left of a city that gave birth to one of the world's first noted philosophers.

Little is known about Thales of Miletus. He was allegedly born in the Ionian city in about 640 BCE to Phoenician parents, making him the son of immigrants. The 3rd-century Greek biographer Diogenes Laërtius quotes the claims of earlier historians that Thales's parents were

Phoenician and he lived into his late seventies. It seems he had a taste for geometry, a willingness to help his king fight the Persians, a need to figure out what stuff is made from and a dislike of children (or at least according to one of Diogenes's accounts, a reluctance to ever get married and have his own).

Rather than spend his nights at the local singles bar, this childless bachelor whiled away the hours pondering if there was anything simpler than water. This might not seem all that unusual for a Greek philosopher; however, Thales deserves a special mention simply because if anybody had asked such a question before, history has since forgotten them.

Philosophy was a fundamentally new form of storytelling, one that attempted to describe nature according to universal rules rather than the wills and emotions of a powerful being. Mathematics and geometry weren't exactly new concepts. Neither was astronomy. People had been recording and measuring the world long before Thales earned fame thanks to the historian Herodotus, who recounted his prediction that the sun would disappear behind the moon on May 28, 585 BCE. What differed was the nature of the questions Thales asked about these measurements.

The radical idea Thales embraced was that all things could be described by something more fundamental. More precisely, just as cakes are made of eggs, sugar and flour, most of what we see is a mixture of simpler materials. Thales wondered what eggs, sugar and flour were made from. And, in turn, what made those simpler ingredients . . . and so on. Thales reasoned that the simplest ingredient of all had to be water. From this clear, mundane liquid all other things could be made.

It's unlikely that Thales spontaneously dreamed up this question on his own one morning. The 5th-century Greek philosopher Plato—famous for being a student of Socrates and for writing the book *Republic*—mentions him in a list with six other wise men in his text *Protagoras*, satirizing their tendency to address complex issues in life with short phrases like "Know thyself." It's therefore likely there would have been discussions between Thales and his local peers at social gatherings, bouncing from one topic to the next, inspiring imaginations to speculate on the relationships between their observations.

Thales's question provides us with a glimpse into a culture of inquiry that was slowly developing on the coast of Ionia during this period. Describing aspects of nature using consistent, impersonal laws was an important step in moving away from the mythologies once used to explain natural events. Around the same time, philosophers in the East such as Confucius and Lao Tzu were writing on human behavior and morals. However, their form of philosophy did not attempt to describe the universe impersonally, but rather focused on improving the individual's role in their community. Why can't we find similar ideas to the Greeks' recorded in other places in the world at that time? Where are the non-Greek natural philosophers? After all, people all over the world had been settled in communities for thousands of years. They could count, write, build impressive structures from stone, smelt metal, sail the oceans and record the movements of the moon and other planets.

A clue might be found by asking what made Greek society on the Turkish coast so different from other Mediterranean and Mesopotamian communities during the 6th century BCE.

The lands surrounding the Nile, the Tigris and the Euphrates were lush and fertile. Life could be rather comfortable for those who spent their days working the soil and growing crops. There was comparatively less to be earned by leaving the community for a major city, let alone going abroad to a foreign country.

Greece, on the other hand, wasn't so fortunate. Its steep, stony hills, small rivers and thin rainfall weren't suitable for cultivating the most common forms of agriculture. Many ancient Greeks figured it was better to gather the relatively few resources they had, such as olives and grapes, and set sail to other lands in search of trade opportunities. This created a widely dispersed community of related travelers rather than isolated clusters of people who grew everything they needed and rarely ventured very far from home.

For our distant ancestors traversing the Paleolithic landscape, encounters with complete strangers were rare. Even if you weren't personally acquainted with a new face, there was a good chance that you would share enough language to communicate and hold relatively similar beliefs with those in your region. Trading further afield encouraged people from radically different backgrounds to communicate with one another, leading to the exchange of fresh stories and ideas as readily as unique goods and services. From this cultural milieu some people would undoubtedly have been presented with worldviews that forced them to confront their own.

It's a neat idea that could explain the rise of the first philosophers, if not for the fact that the Greeks weren't the first intercultural traders. Phoenician sailors had ruled the seas for generations, trading in dyes, glass and timber, and possessing an advanced system for writing and mathematics. In fact, the Phoenician merchant vessels explored far further abroad than the Greeks ever did, leaving the Mediterranean and touching on the shores of the British Isles far to the north. Encountering different mythologies would surely have inspired some Phoenicians to compare their beliefs with others, impacting on the stories they told. History is full of examples where stories merged and evolved under the influence of a new culture. To lead to a radical new epistemology, however, these societies needed more than just a conflict of opinions to seed the first philosophical thoughts. If good foreign relations were solely responsible for challenging mythology as a way of seeing the world, why didn't Phoenicia initiate such a philosophical revolution?

Throughout its period of global dominance the Phoenician civilization had been governed by a monarchy. The aristocracy were also merchants, so most of the civilization's trading wealth remained tightly under the control of the ruling class.

No single factor on its own can be held accountable. There are, however, several features of Greek society that, when combined, just might have been enough to fan the flames of a new way of thinking about the fundamental mechanics of the universe.

Historian and mythographer Henri Frankfort describes differences between the Greeks' relationship with their gods and those of contemporary cultures, suggesting a weakening in their gods' omnipotence and an absence of strong religious dogma might have made way for a more objective description of the cosmos. On the other hand, the German philosopher Eduard Zeller believed it was both the birth of the republican system of government and the expansion of Greek colonies that precipitated the advance of philosophical thinking, claiming it to be a "condition of freedom."

By late in the 6th century BCE, the Greeks had entered the age of democracy. Emerging from the social conflict created by an aristocratic system of rule in Athens, the laws written by the statesman Solon to mediate factional rivalry led to the foundation of a more democratic form of governance. Therefore, power—and, importantly, wealth—could be earned through a means other than title and inheritance. With the relatively recent invention of coinage in Lydia, a more fluid system of economics was born, where wealth could accumulate without going rotten or taking up a warehouse, be borrowed in large sums and divided into small, finite amounts.

The cities on the coast of Ionia, such as Miletus, Pergamum, Smyrna and Ephesus, were famous for being trade ports. Individually wealthy merchants raised in a democratic society might have had more freedom to contemplate the mysteries of the world, where previously it simply wasn't economical to spend time on such frivolous academic pursuits. Artists and thinkers could have wealthy patrons to commission their works. Significantly, an economy could permit a greater level of social distance from the traditional community. Where myths were useful for binding a tribe together, they weren't quite so important to rich trading communities who were relatively free to pick and choose where their next meal came from, thanks to the value of the coin.

Whatever the underlying factors were, cultural freedom to think critically undoubtedly played a key role. Being philosophical means more than just asking questions like "what is stuff made out of?"—it allows an individual to find fault with the beliefs of those they would otherwise be reliant on for food and protection. Once upon a time, such disagreements were unthinkable. These social changes made it possible to not only contemplate an impersonal universe, but to form an academic culture where one could argue about it without fear of repercussion. Gods and goddesses weren't completely done away with, but their role in explaining natural phenomena was diminished. Alliances with supernatural personalities weren't as necessary, making impersonal laws more practical in describing events.

There is no doubt that the philosophical revolution in ancient Greece produced a novel way of thinking that conflicted with traditional mythology. Drawing a path between Thales and the complex beast that is modern-day "science" is arguably easy to do, at least in hindsight. But somewhere along history's path, our way of thinking crossed a vague boundary and went from merely contemplative to "scientific" in a modern sense.

But what does it even mean to call something "scientific," as opposed to just "philosophical"? Are all philosophers intrinsically scientists? Identifying the features of science demands a closer look at the word itself and an exploration of its many contexts.

THE NO-MAN'S-LAND OF SCIENCE

For as long as I can remember, I've loved to draw, especially cartoons. I'm a great fan of art. I'll happily spend an entire day walking through a gallery of portraits, or sitting on my couch, flicking through a graphic novel. A well-crafted sculpture can capture my attention just as easily, not to mention a night out at the theater, a good movie or a music concert.

But not all art pieces impress me. On occasion I'll walk past some monstrous collection of glass fragments and twisted metal and wonder why somebody would proudly put it on public display. I might force myself to tolerate a performance piece, stunned at how tragically awful it was, or endure a ghastly cacophony of noise and wonder how anybody could call it "music."

"That's not art!" I'd say, blowing a dismissive raspberry.

My wife often disagrees. Regardless of whether she likes it or not, she'll insist it's deserving of being called "art" even as I roll my eyes and sigh obnoxiously.

It's futile trying to convince her. Not because she'd stubbornly stick to her guns, but because neither of us can prove that our perspective is the correct one. There are certainly objects we agree constitute art, just as there are many objects or situations we agree could never possibly be described that way.

However, there is a place in between where our different opinions make it difficult to draw a clear line. For instance, I believe that art is defined purely by the artist's intention, while my wife feels it necessarily involves the viewer's appreciation. I could find knowledgeable people who agree with me, but so what? I'd imagine she'd have no trouble doing the same. I might even find that there are more people who agree with her, but are meanings always democratically elected, or do they simply rely on the barest patches of common ground?

We can think of this as a "demarcation problem": a place where the meanings we associate with words reveal differences in understanding imposed by the wide variety of personal experiences we all have. My tribe of agreeable comrades might think art is about the intention, while my wife's tribe understands the word inherently implies a feature of interpretation. Each of our communities might use the same word— or, in terms of intentionality, the "signifier"—to describe two subtly different concepts—or "signified"—introducing confusion in the process as we attempt to communicate.

Like art, science also has a demarcation problem.

Everybody has an opinion on what science is and what it isn't. The 20th century saw countless attempts by modern philosophers to embed the topic within a concise definition or a simple qualifier that is meant to be the final word on the matter. None has been universally accepted, and it's unlikely we'll ever have a simple dictionary definition that can be constrained within a paragraph or two. It seems there are as many meanings as there are people.

The word itself comes from the Latin scientia, which simply means "knowledge." Yet how many people would describe the fact they "know" their street address as scientific? Is my name readily described as a scientific fact?

That said, some types of knowledge are often described under the banner of science. The distance from the Earth to the Sun can be found in many school textbooks labeled as a scientific fact, as can the chemical formula for water. Why would I sooner think *Canis familiaris* is a scientific name, while the common word "dog" isn't? What makes that knowledge different from my name and street address?

Another definition portrays science as an activity people do. The 19th-century English philosopher, priest and historian William Whewell was the first to coin the term "scientist" to mean "natural philosopher" as he attempted to systemize the history of natural discovery in his 1840 book *The Philosophy of the Inductive Sciences*. Just as artists make art, Whewell figured scientists make science, claiming, "We need very much a name to describe a cultivator of science in general."

Science is also frequently described as the principles informing the activity: a method that starts with a question phrased as a statement, or a "hypothesis," which progresses to being tested against observations in a contrived, experimental environment, and eventually ends in a definitive conclusion. But history commonly recognizes people as scientists (and even praises them for their contributions to scientific knowledge) who haven't strictly followed such a method. Physics is full of amazing discoveries that have resulted not so much from experimentation but from mathematical deductions, or even numerous hours of patiently watching the sky or perhaps sifting through photographs. The discovery of phenomena such as black holes or subatomic particles is often seeded in extrapolations that we typically think of as scientific, long before they are found or identified by an experiment under controlled conditions. While evolution can be demonstrated in the laboratory today, Darwin had no such luxury and relied on comparisons he could make between diverse environments.

Then there is science described as technology. Everything from computers to cloned sheep is commonly labeled as "the latest science." A new mathematical formula, a planet discovered outside our solar system, a genetically engineered plant . . . innovation and discovery are regularly emphasized beneath a heading of science in the daily newspaper and the evening news. "Frankenfoods" are demonized as bad science while the cure for cancer is good science. Strangely, movies typically refer to the makers of death rays, hybrid animals and body-altering drugs as mad scientists rather than mad engineers. Science is sometimes in the product, it seems, rather than the method.

Was the first chimpanzee to modify a twig and poke it into a termite nest a scientist, then? What of the first spear maker? How about the first human to create fire? They are certainly comparable to any modern form of technology, at least as a progression of tools. But is it fair to say we've been doing science long before we ever stumbled across the philosophical processes that allowed us to model ideas and question them critically before putting them into practice? Is the trial and error method the same thing as science?

Science is undoubtedly a contextual term that is hard to define on its own. It might appear to be rather pedantic to try to squeeze it into a tightly held definition in light of such a broad field of meaning, yet we often use the word in isolation without giving any thought to what we mean. Governments are becoming aware of the need for scientific literacy among their citizens, for instance. But does this mean people should be aware of all new and imminent technological advances? Should the public understand what it is different scientists do? Does scientific literacy refer to a grasp of critical thinking and skepticism? Does it imply that people should know more scientific facts? Are all of these things of equal importance? Faced with limited resources and a quest for finding the most efficient ways to educate people about science, breaking science down into distinct contexts is more than just a philosophical exercise.

Knowing what it means precisely for an idea to be scientific, as opposed to mythological (or purely speculative), is rather important. It could make the difference between a building staying up and falling down, or determine whether you choose medicine that makes you better or medicine that kills you. While a demarcation problem in art might result in a heated discussion, it hardly matters if a person purchases a canvas of spilled paint believing that it is art. In science, this vague no-man's-land is far from a trivial matter.

Nailing down the precise features that define science using little more than common usage and historical dates is an impossible task. There is a long list of philosophers dating back over the centuries who have tried (and failed) to come up with a strict set of terms that we can all agree on. Yet somewhere amidst the variations in connotations, colloquialisms and contexts, there is enough overlap to find a starting point.

A SCIENTIFIC SEQUENCE

Ibn al-Haytham (also known as "Alhazan") was born in Basra around 965 CE, and spent his life studying everything from anatomy to engineering to optics, contributing a vast range of discoveries to our understanding of astronomical physics, which made Newton look like a copycat. As with most Arab philosophers, he shared the same belief as his Greek predecessors that human perception is innately flawed and incapable of knowing reality with absolute certainty. Unfortunately, again like the Greeks of old, he also believed that we couldn't do much more than use our senses in combination with our talent for reason, in spite of their intrinsic failings.

Al-Haytham advocated the rigorous application of a specific process to help create a sense of consistency between accounts of different observations. In his book on optics, *Kitab al-Manazir*, he demonstrates a single, preferred method for actively testing ideas:

1. Observation
2. Statement of a problem
3. Formation of a hypothesis
4. Experimental testing of the hypothesis
5. Analysis of the results
6. Interpretation of the analysis/formation of a conclusion
7. Dissemination of the findings

While rigorous experimentation itself wasn't novel, there's no earlier record of such a detailed, step-by-step method. Al-Haytham also understood the role doubt played in evaluating his ideas, stating in an essay, "Therefore, the seeker after the truth is not one who studies the writings of the ancients . . . but rather the one who suspects his faith in them and questions what he gathers from them."

Al-Haytham's belief that everybody was prone to making errors of judgment was common among the early Greek and later Arab natural philosophers. Without a detailed understanding of how the brain functioned, they were still aware of its limitations and how it was open to imbuing illusions with unwarranted meaning. Logically, competing ideas that disagreed with one another couldn't both be equally correct, meaning perception had to be wrong on at least one account. Unlike mythology—where consistency wasn't important—natural philosophers believed there could only be a single correct answer.

If we are all capable of making mistakes, how do we determine which ideas are likely to be "right"? Is this even possible, or should we all take the advice of the ancient skeptic Pyrrho of Elis and simply find peace in our ignorance, given we can never definitively know anything with certainty?

The 4th-century BCE Greek philosopher Aristotle described in his text *The Organon* a system of thinking based on a formulaic arrangement of concepts. By comparing accepted premises it was possible to infer a novel conclusion. As conclusions were accepted, they could also be used as premises in other logical forms, allowing for new knowledge to be created without being directly observed. This process of "deductive syllogism" contributed greatly to the study of nature by providing self-evident principles from which to work, yet by the dawn of the Renaissance it was felt to be insufficient in solving many of nature's mysteries. Aristotle's conclusions on the natural world had slowly fallen due to experiment and observation, demonstrating a weakness in his approach.

The Elizabethan Lord Chancellor and Attorney-General Sir Francis Bacon wasn't responsible for any new scientific method or amazing discoveries; he simply summarized the efforts of philosophers throughout history to come up with a single explanation of what they were doing and, in his humble opinion, how they should be doing it. Bacon proposed a similar method to al-Haytham's for comparing ideas.

Bacon was critical of the philosophers who had been praised throughout the Middle Ages, such as Plato and Aristotle. Bacon believed the time was right to develop a new form of reasoning which was better than the old deductive syllogism system his contemporaries dogmatically adhered to. He didn't reject all of Aristotle's work outright, but criticized the vague, airy-fairy terms that he believed were responsible for philosophers' reluctance to let go of obsolete or incorrect beliefs regarding natural phenomena.

Bacon recognized that human perception was rife with prejudice, which could easily interfere with the process of investigation. He called these biases "idols," each reflecting the notion that human thinking is programmed to deal with other people as opposed to an impersonal universe.

Bacon's Idols

1. Idols of the Tribe—Our tendency to overgeneralize and focus only on those observations that support what we already believe.
2. Idols of the Cave—Our tendency to subscribe to the beliefs of our social group.
3. Idols of the Marketplace—The limitations of our use of language, which make it difficult to describe or communicate some ideas because words aren't specific enough.

4. Idols of the Theater—Our tendency to readily accept the views of an authority for authority's sake.

Bacon no more invented science than Thales invented philosophy. Even today, what is commonly described as the scientific method bears only a passing similarity to Bacon's idols, tables and method, or indeed al-Haytham's experimental process. Yet by seeking ways to limit the impact of our social way of thinking on our perceptions, philosophers like Bacon created a set of measures against which ideas could be compared. The best ideas proved themselves to be useful by successfully predicting the future, providing hints that increased the chances of a new discovery or led to useful innovations, while the worst ones conflicted with observations and led to more complicated questions than they solved.

Therefore, while science can be many things, above all it is a way of using our mistake-making, illusion-prone, storytelling brains to compare different methods for describing nature.

Nobody would say there is only one way to make a cake. But most people would agree that good cooking should endeavor to make food that suits certain expectations, such as nice aesthetics and a pleasing taste. My mud pies might resemble cooking, but I won't be opening a restaurant with them any time soon. Likewise, science isn't a recipe, or even a good meal. It is not a single method, a list of facts or a community of people any more than cooking is distinguished by a single recipe or a union of chefs. Rather, science is the discussion we have about whether a belief reflects values that are likely to make it useful in describing the universe, just as we can discuss why certain ingredients and implements make for a delectable chocolate gateau. In either case, the proof of the pudding (as they say) is in the tasting.

THE PHILOSOPHER'S TOOLBOX

During the 18th century, Mesmeromania took hold of France. Franz Friedrich Anton Mesmer was a German physician whose notoriety could compare easily to any of today's A-list celebrities. Although witty and charismatic, it was his alleged ability to heal with a touch that won him such adoration from the French aristocracy.

It was commonly assumed during Mesmer's era that a relationship existed between the movement of celestial objects and health—in fact, the term "influenza" hails from the mistaken understanding that the shifting heavens could influence illness. Mesmer's medical dissertation concerned the effect the moon and planets had on disease as a result of Newton's relatively novel gravitational force.

Drawing on similarities between gravity and the pull of magnets, Mesmer experimented with magnets and iron to induce "artificial tides" in his patients, echoing the earlier work of Jesuit priest and astronomer Maximilian Hell, who had also dabbled in magnetic healing practices. Mesmer found that he could achieve the same effects as the priest without relying on any such metal accoutrements, and quickly did away with them altogether.

He came to call the phenomenon gravitus animalus, or "animal magnetism," a term that has evolved to mean something very different. In this context it referred to a mysterious, invisible force that permeated all things. When it concerned living objects, it was an "animal" force, from the Latin animus, meaning breath or spirit. Mesmer claimed he could influence the flow of this ubiquitous quintessence with his own animal force; a handy talent given that any impediment to the flow of this mystical force through an organism was believed to give rise to illness.

The physician's fame spread, and, as often happens to those in glamorous positions, rumors followed of his involvement in less than respectable pastimes. If tabloids had existed back then, Mesmer would have been on the front cover with the headline "Queen Mesmerized by Shocking Affair." In the hedonistic culture of 18th-century upper-class Paris, it's easy to imagine that a charming, popular gentleman suspected of indulging in rather sordid, late-night parties might inadvertently tread on the wrong toes. Mesmer's close friendship with the queen, Marie Antoinette, caused her husband, King Louis XVI, great concern. Unfortunately, a direct accusation of any unsavoury acts would risk making the queen look bad, leaving the king to find another way to discredit the healer.

Accompanying Mesmer's fame was a more critical form of attention—not all of his patients were impressed with his methods, while many of his colleagues and fellow academics were suspicious that there wasn't anything to his claims. They wrote to the king and expressed their doubts, prompting him to gather some notable minds to investigate.

Among them was the recently appointed American ambassador, Benjamin Franklin. He was joined by statesman and chemist Antoine Lavoisier, the mayor (and astronomer) Jean Bailly and the inventor of the guillotine, Dr Joseph Guillotin.

The immense public support for Mesmer's claims came thanks to the testimonies of his patients. Sufferers of all sorts of illnesses and ailments, from a range of backgrounds, claimed to have felt much better after his sessions. In fact, such was his popularity he altered his methods to work on groups and employed students to tour his unusual craft. For Mesmer, the proof was self-evident—his methods were obviously fruitful. It was ludicrous that they would need to be "investigated," and he was insulted that he should be the subject of a royal inquiry. He refused to cooperate; his only concession was sending a student—Dr Charles D'Eslon—in his stead. It was a sneaky strategy: if the student were successful, it would be proof positive of the effectiveness of mesmerism. If he failed, it would only speak of his incompetency and not Mesmer's.

As Franklin suffered heavily from gout, he was reluctant to travel from his residence in Passy. His condition made him extremely open-minded (and even hopeful), as did his son's conviction in the healer's abilities as well as his own friendship with the queen. Yet Franklin also knew how easy it was for somebody to be fooled out of desire. The only answer lay in testing these claims personally.

D'Eslon went to Passy with a selection of his patients. The commission witnessed remarkable scenes, ranging from people collapsing into trance-like states to people dramatically coughing and spitting in violent convulsions. The displays seemed to be contagious, with the rest of the crowd quickly falling into similar fits. While unusual, it seemed obvious that D'Eslon held some sort of control over his patients. Franklin, however, felt nothing. In spite of several sessions, he experienced no such trance, convulsions or even an alleviation of pain in his feet. However, this was not in itself evidence that the claims were false—it was simply that they were not effective in his particular case.

An alleged feature of Mesmerism was the ability to "charge" an object in some manner, allowing patients to rely on it for treatment should there be no practitioner handy. Franklin saw this as a way of testing the claim without necessarily experiencing it himself. If he could see this object work on other patients, it would support the conclusion that there could be something to this claim of animal magnetism.

There was a slight problem. Franklin and his colleagues knew that the patients expected this method to work. After all, they were all selected by D'Eslon and were convinced of his skills. To get around this, the investigators decided that it was important to abstain from revealing which object was "magnetized," leaving the patients to decide based purely on their experiences.

This must have been deemed suitable by D'Eslon, who proceeded to magnetize a tree within a small wood without being observed by the patients. A 12-year-old boy was then brought forward, blindfolded, and asked to determine which tree had been magnetized. After standing about, presumably scratching his head in confusion, he eventually made his selection. Needless to say, it wasn't the right one.

This was the first recorded instance of a blinded test, where the variables in an experiment were hidden from the subject. Investigators now had a powerful new tool that could be employed in such situations to identify ideas that were purely the result of a subject's biases.

Of course, it couldn't prove with absolute certainty that Mesmer was wrong. It simply provided the commission with another observation that might help them identify which explanations for his apparent "talent" were more likely to be true. Mesmer himself, because of his expectations, was drawn to an explanation that involved mystical therapeutic talents. Without the means, or desire, to find contradictory observations, he was left with just the one explanation: animal magnetism. While there are those who maintain Mesmer was a clever charlatan, few of his time seemed to doubt his sincerity.

Philosophers have long understood that it is impossible to discern with absolute certainty ideas that are certainly true from those which are false. However, over the centuries numerous tools have been developed that serve as a guide to indicate how confident we can be in deciding that one explanation is more or less useful than another.

Science is like a game of "Who Wants to Be a Millionaire?" where you have no idea what the correct answer is. Before making a guess, you can phone a friend, ask for a 50–50 choice or poll the audience for their opinion. If you do all three, you can increase your confidence before you pick an answer.

Fortunately, there are more than three ways of increasing your confidence when it comes to science.

SHADOWS ON A WALL

Our senses are the only conduits between our mind and the universe. The philosophical position of "solipsism" maintains that because it's impossible for any individual to step outside of their mind, it cannot be demonstrated that there is anything external to our own thoughts. What we appear to sense cannot be distinguished from what's in our mind. Even if it's true and the universe exists only in my mind, I'm still left with the need to explain a system that appears to operate external to my thinking about it.

In his book *The Republic*, Plato has a character explain an imaginary scene in a discussion on education. The scene presents the reader with a row of prisoners sitting chained to the floor of a cave. Shackles prevent them from turning around, forcing them to stare at a wall of stone. Behind them roars a great fire. Between the bright flames and the prisoners' backs is a walkway, along which a line of figures marches. The prisoners watch this endless parade of shadows—black, featureless shapes rippling across the rock in single file. The allegory is Plato's way of pointing out the difference between our experiences of the world and their ultimate causes.

Observation can be considered to be the most primary of the philosopher's tools. Our universe is represented in terms of sight, sound, scents, tastes and tactile sensations. On top of that are our physiological responses—emotions that drive our reaction to these stimuli. Distinguishing where one ends and the other begins is often difficult to do, leading us to conflate our emotions with our observations. Through our ability to empathize and communicate, we can embed our observations and our emotions into symbolic thought and try to transfer them to another individual, sharing our experiences.

"Empiricism" describes "knowledge as defined by experience." At its very heart, science is concerned about describing the universe as it appears to our senses. It's our experience of the world that prompts us to ask, "Why this and not something else?"

Of course, the problem is that "cause" is a conclusion we come to, not a property we directly observe. Thunder might follow a flash of lightning, but, speaking empirically, we can only say we saw a flash of light and heard a boom in a timely sequence. We can experience the sensation of a touch, observe a tumbling glass and hear it smash on the tiles, and describe the sequence of events as "I bumped the glass and knocked it to the floor." Few would argue that the acts aren't related; however, a "cause" is not something we can see—it is a property we ascribe to a sequence of events based on their relationship in time and space.

Such correlations are so innate to our brain we find it difficult to avoid treating correlation and causation as if they are one and the same. Spying a black cat only to experience bad luck moments later immediately becomes suspicious as a causative relationship, rather than innocently correlative. Again, it's our brain making a reasonable guess to save time and energy. Causes are often correlated, and most errors we make are of small consequence when compared with the numerous times we get it right.

For all of its faults, empiricism is still a most useful philosophical tool. As far as our brain is concerned, there is nothing but the experience of sensations and its ability to combine them in our imagination. On its own, however, empiricism limits us to doing little more than painting a static scene of nature. As in Plato's cave, we must use more than the shadows on the wall; they must be combined reasonably with the smell of the fire, the flicker of its light and the sound of rustling feet echoing from the walls to imagine a scene that guesses at what lies beyond what we can observe.

THE SHARPEST TOOL IN THE BOX

Our ability to combine sensations in our mind in ways we have never experienced makes imagination a powerful tool. There is virtually no limit to the ways we can picture how observations might be related; most children are incredibly adept at stitching together complicated fantasies from the barest scraps of experience. It's easy to be overwhelmed with possibilities, where literally anything that can be imagined is a potentially useful belief. Therefore, it's important to determine which scenarios are most worthy of further consideration and which are unlikely to be productive.

The Franciscan friar named William of Ockham initially seems like an unlikely source for one of science's most often used tools, given his dogmatic faith. From him we get the maxim of "Ockham's Razor"—a principle that is often misunderstood yet vitally important to natural philosophy.

The philosopher Plato maintained that for all things there was a perfect archetype. This flawless form wasn't thought to exist in a way that could be experienced. It was described as "metaphysical." As such, while we experience square-shaped objects, the archetype "square" exists as a universal form. This was more than just a property or a description—it could be said to exist in real terms. It just wasn't an actual "thing" in the same way dogs, watches, stars, carrots and lumps of rock are "things."

Ockham disagreed. He argued that only those things that could be directly observed could be said to be real. In other words, where Plato suggested that beyond all actual dogs there was some perfect yet intangible form of "dog," Ockham would have believed that there were only the dogs we see walking around the streets; anything more abstract existed only as a figment of our imagination.

Since Ockham was a friar, he believed the only concept in the universe that could not be directly experienced but still accepted as real was God. Ockham referred to this using the phrase "*numquam ponenda est pluralitas sine necessitate*," which roughly translates into "nothing should be multiplied beyond necessity." The philosophical term is "ontological parsimony"; however, today Ockham's Razor is used more as a rule of thumb rather than a strict universal law. The rule suggests that in searching for an explanation, we shouldn't be confident about an idea unless it is self-evident or able to be directly experienced.

Sometimes this rule is explained as "the simplest of multiple explanations are usually the best." Indeed, throughout history philosophers have remarked on the simplicity of nature. The 13th-century priest Thomas Aquinas phrased it as, "If a thing can be done adequately by means of one, it is superfluous to do it by means of several; for we observe that nature does not employ two instruments where one suffices." However, the problem with this definition of ontological parsimony is that simplicity is a difficult quality to ascribe to an idea, at least objectively speaking. What seems to be simple can hide a great many complexities.

A better way of understanding this maxim is to see it as an economy of assumptions. Assumptions are little more than rough guesses that inform our choice of belief, often guided by our desires rather than reason or observation. As ideas grow needlessly in complication, they risk being supported by a greater number of assumptions. The more wild guesses an idea requires to be considered true, the more opportunity there is to be wrong. Therefore, ideas that rely on the fewest assumptions have the greatest chance of being useful as a belief.

A TEST OF STRENGTH

In 1971, Apollo 15 astronaut David Scott stood on the surface of the moon with a hammer held aloft in his right hand and a falcon feather in his left. He announced that he would test Galileo's prediction on falling masses and let both objects go. Needless to say, they hit the moon's powdery regolith at the same time.

Nobody doubted this would happen. Similar demonstrations using towers, ramps, ball bearings and vacuum tubes had been repeated time and time again over the centuries. Galileo was said to have dropped weights from the tower of Pisa. If he did, he never left us with anything more than a few words outlining his thoughts on the matter. But he did set out to disagree with Aristotle's intuition.

Aristotle stated that heavy objects fall faster than light objects.

Given that leaves tumble lazily from a tree's branches while its fruit drops quickly, he could almost be forgiven for thinking that.

Aristotle wasn't one to do experiments. If he had done, he would have had good reason to question his assumptions. Then again, if he had lived long enough, it's possible he might have one day accidentally stumbled across a situation that contradicted his belief. He might have watched two rocks tumble from a cliff—one big and one small—only to see them splash into the water at the same time. Experimentation meant he could have actively created the situation for that observation to occur, rather than stumbling across it by chance. He also could have controlled the situation carefully by weighing the rocks, making certain that he knew as accurately as possible what was going on.

There are many things we cannot do in a laboratory. We can't put a solar system into a box or re-evolve birds from dinosaurs in a glass tank. Yet even though we can't directly manipulate some variables under clinical conditions, we can still compare observations in similar experiments that occur naturally.

Our ability to physically eliminate confounding variables is singly the biggest advantage of employing an experiment over passive observation. While experimentation had become common practice by the 17th century, naturalist Francesco Redi is regarded as the first to actively conduct trials which controlled for variables, which he published in 1668 under the title *Experiments on the Generation of Insects*.

The question of "spontaneous generation" was being critically debated during Redi's time. Could life emerge from conditions where none previously existed? Could moldy straw produce baby mice, or were parents needed? Spontaneous generation was a notion that was favored by those who believed a vital essence permeated the universe, giving rise to life out of chemical ingredients. What we take for granted today was a perplexing mystery several centuries ago. Redi formulated a way of testing the idea that rotting meat could spontaneously produce a swarm of maggots.

Rather than simply leave a chunk of meat out on the bench for a few days and watch it intensely, he created two situations. Both constituted three jars of different varieties of meat. One set he covered with gauze and the other he left open, thus controlling for the variable of access for adult flies. Given that the open jars produced maggots while the gauze-covered jars of meat did not, he had provided evidence that life required a parent generation to form.

Today, experiments are even designed in ways to permit observations of events that would disprove a hypothesis. The large hadron collider in Switzerland, for example, smashes subatomic particles together so physicists can examine the pieces. Paradoxically, one of their goals is to locate a particle nobody believes exists—an excited quark. For as long as nobody sees one, its absence demonstrates that the quark isn't made of smaller particles. Spotting one in an excited state would indicate this is not the case, forever changing our understanding of fundamental particles. It is the fruitless search that provides evidence in such a case—an experiment designed to prove an idea wrong.

WHAT ARE THE CHANCES?

You've just sat down to a friendly game of cards with one other person and played a couple of rounds. They've beaten you twice, and you've just won the last hand. As you shuffle in expectation of another game the fire alarm rings and you have to evacuate the building. There's $50 riding on the outcome, which was going to be given to the first person to win ten games. What's the fairest way to divide the prize?

Gambling on games of chance is one of the world's oldest thrills. Yet "fairness" isn't a simple concept to apply, especially when the game suddenly changes. In the 15th century, the Italian mathematician Luca Pacioli suggested the fairest way of dealing with a premature end to the game was to divide the pot between the players according to how many games they'd already won. This is all well and good, but what if the match was interrupted after just a single round? That player wins it all . . . hardly fair when the game hadn't even warmed up.

Half a century later another Italian mathematician named Niccolo Tartaglia introduced the idea of taking into consideration the number of rounds needed for a player to win. Hence, it was a matter of combining the winning player's lead, number of hands won and number of rounds left to play. However, he still wasn't happy with this answer, and felt it was ultimately subject to the agreement between the players anyway.

It was this question of what constituted fairness in game-play that ultimately led to the creation of the mathematic field of probability. In the mid-17th century, French mathematicians Blaise Pascal and Pierre de Fermat sat together in a café and chewed on this age-old problem in search of a better answer. The most important thing, they decided, wasn't the number of rounds a player had won, but how many rounds they still needed to win the game. The trick was to count out all the different ways either player could achieve that number of wins and calculate which player had the best chance of success, given their current score.

Over the following century, variations of Pascal and Fermat's formula were applied to phenomena other than gambling. For instance, the astronomer Christian Huygens was encouraged by Pascal to write a book on the topic, titled *On Reasoning in Games of Chance*. A significant principle on the margin of positive and negative errors in calculations was later developed in the mid-18th century, allowing people to determine how confident they should be that a specific answer to a problem was the correct one.

Probability theory deals with using known quantities to determine the likelihood of an unknown event occurring. While people had long collected numbers and recorded measurements of the world, there were few mathematical tools to combine them in ways to fill in the missing pieces. With Pascal and Fermat's discussion came the inspiration to examine numbers in a way that would describe apparent randomness. We might not know which number will come up on the roll of a dice, but we can describe all of the events that could happen, and compare their likelihood of rolling face-up.

As thinking tools, probability and statistics provide us with the means to better describe randomness. Rather than face the unknown of illness, epidemiologists can provide us with a picture of how likely it is we'll fall sick from a particular pathogen, such as the seasonal flu. Immunologists can give us an indication of the chance a vaccination will work as planned or have unwelcome side effects. Combining the figures allows us to compare probabilities and outcomes and determine whether it's a greater risk to get a shot or to leave it and possibly get sick.

Accurately measuring chance is not an innate skill we humans possess, and is troubled by a range of biases. We find it difficult to intu-itively know where to draw the line around a set of related events and allocate properties that describe their total frequency. A trip to the local casino is all it takes to demonstrate how we place too much con-fidence in our poor probability skills. Our desperate need for certainty, as a result of our brain's more exuberantly rational features, means we find it difficult to judge scales of "maybe" in contrast with the absolutes of our actions. We eagerly seek patterns that will assist us in our judg-ments, sharing knowledge that might further help us predict the likeli-hood of a conclusion. Usually, this knowledge exchange is productive. Sometimes it isn't. Yet with the tools of probability, we are better adapted to quantify and measure a universe full of random unknowns.

THE PHILOSOPHER'S BURDEN

Drawing a line between philosophy and science is a fool's game. There was no definitive point where the natural philosophers of the ancient world could be said to have become scientists. Philosophy simply gath-ered more tools with which we could gain more confidence in an idea being useful to describe nature.

In many ways, that first principle of science—to describe the objec-tive laws that seem to govern the universe—hasn't changed since Thales. The tools, however, have diversified and become increasingly complicated. Experiments now demand the use of positive and nega-tive controls. Surveys can pick different types of information from the perceptions of many individuals. Statistical tools can be modified to take into account different types of data. Tests are often double blinded to remove not only the bias of the subject, but that of the experimenter as well. Scientists today have the choice of discussing their results with the public through not only journals but other media sources such as television and the Internet.

Not all tools might be suitable for all questions. The choices faced by scientists need to take into account risk, cost, time and reward. Ethics need to be considered, preventing us from conducting experi-ments of various sorts on people or animals that might conflict with our moral values. Some logical deductions might be directly unobservable, forcing us to find other ways of perceiving them. Some qualities don't lend themselves to being quantified easily, losing meaning in the process of being calculated as a statistic.

Science is the discussion of whether an idea is merited, aided by the correct use of tools from the philosopher's toolbox. As with all discussions, not everybody will agree. Opinions will vary with the culture of the scientist, their own experiences and their own biases.

In many ways, science has not just come to replace our myth-making traditions; our storytelling heritage has adopted science as a theme in its own right. Science fiction has become a popular genre in the arts, so too "edutainment" in the form of documentaries and popular science media. But is it science, or a mere simulacrum? The edge of the discussion is occupied by a shadowy borderland, a place where it's easy to get lost and have too little or too much confidence in an idea simply because it feels scientific. It's a place where the social brain takes control and objectivity falters, where we attribute the features of critical inquiry without applying any of the tools. The intrepid traveler believes they are still doing science and yet have wandered far from its fertile ground.

It's easy to see why this happens. For millions of years we've looked up at the stars and seen the shapes of gods, animals and people. For only the past few hundred years we've looked up at the stars and seen other suns. Today we see the orbits of planets like our own. Beyond them we can even see light from the edges of time, twisted and deformed by the accelerating expanse of space—almost a magical, impossible concept in itself.

Sadly, because of our tribal brains, science carries a hefty cost. Treasured ideas that are loved by the community may be left behind, unable to compete with conflicting observations. Admired heroes may be found to have been mistaken. Years of hard work can amount to nothing thanks to a single observation, making a lifetime of effort seem like a waste of time. For our tribal brain, the philosopher's toolbox is full of double-edged knives, capable of cutting away our hopes with the myths.

Chapter 3

THE PITIFUL MONSTER

Why do doctors wear white coats?

Science has a lot in common with *The Strange Case of Dr. Jekyll and Mr. Hyde*. We perceive it as both our salvation and our doom, as amoral and immoral, insightful and unwise. We turn to science when we want confidence in a belief, yet only if we already agree with its premises. In the modern age, painting an idea with a façade of science can make it look appealing. Yet on scratching away the jargon and the lab coats, we'll often find there's nothing but bold assertions and wishful thinking. If only it was as simple as asking a scientist what to believe.

SCIENCE: THE GREATEST SHOW ON EARTH

Stillness falls over the London crowd as a gentleman dressed formally in tails and a gray cravat clears his throat and gingerly picks up a thin black rod. With a flourish he darts the wand back and forth through a silk cloth as if polishing it, almost hypnotizing the audience of wealthy gentlemen, ladies in lace and crinoline and the occasional working-class student who managed to scrimp enough shillings for the entrance fee.

Slowly, his hand descends toward a scattering of shredded paper upon a tabletop, prompting the paper to spring to life, shoot skyward and stick to the presenter's wand. The audience applauds. The presenter proceeds into a discussion of charged surfaces and the movement of special electrical fluids.

For most of the audience, it is the spectacle that lures them to the auditorium rather than the detailed lecture on physics. Never is the word "magic" spoken or any mystery left unsolved. They have all gone to see a scientist present a talk on static electricity. But they are entertained nonetheless, as if by witnessing a demonstration of nature's wonder they have been privileged to observe something truly supernatural, if not plain diabolical.

By the start of the 19th century, a class uprising across the channel in France had taken a threatening turn for the worse. What began as rebellion swiftly became a full-blown war that would engage much of Europe. Nobody had foreseen Napoleon's grab for power, yet before long the conflict's impact on trade forced England to question its own self-sufficiency, turning its focus inward to ask how the country would satisfy its industries' needs now that Europe was trapped behind enemy lines.

One response was rather novel for its time—provide the public with a greater understanding of science. The industrial revolution had already primed the public to accept the usefulness of chemistry and physics in providing practical solutions and improving manufacturing and agricultural processes. Now it was simply a case of fanning those flames.

Studying in academia had once been left to those who could afford not just the books and tuition, but patronage to pay for the rest of life's necessities such as food and a few pints of beer on a Friday night. Nobody studied the arts, humanities or philosophy to become rich. Eager students would often leave home to join a community of likeminded individuals, distancing themselves from the world of crops, construction, manufacture and animal husbandry to enter one of arguing, reading, writing and measuring. Theirs was an alien world far removed from the sweat and toil of everyday life. Without the patronage of a family inheritance or devoting their life to God in a church-sponsored school, it was typically impossible for the average citizen to gain a deep understanding of what it was that philosophers did.

That was all about to change.

One March evening in 1799, members of London's esteemed Royal Society met at their president's house to discuss the funding of an outreach project dedicated to educating Britain's citizens on all matters scientific. The project aimed to bridge the divide between the academic and lay communities and to inspire non-scientists to develop an active interest in the products of scientific discovery.

In addition to the funding provided by various philanthropists and groups such as the Society for Bettering the Conditions and Improving the Comforts of the Poor, 58 rather affluent and influential members of Britain's aristocracy each agreed to pay what was then an astonishing sum of 50 guineas to establish what would become known as the Institution. Later, this somewhat plain and simple title would be expanded to the Royal Institution of Great Britain, and its Mayfair address would house some of the greatest minds in British scientific history, including Sir Humphry Davy (who identified the elements chlorine and iodine), Michael Faraday (famous for his work on electricity and magnetism) and James Dewar (known for liquefying gases).

Members of the Royal Institution would not only perform research for Britain's industrial and agricultural sectors, they would regularly present lectures to the general public. Where an evening at the cinema or a concert might be a lot of fun for us, Friday night was science night for many young 19th-century socialites. Scientists quickly became celebrities on the high streets of London akin to any poet, musician or actor. Sir Humphry Davy's golden tongue matched his talent for conducting experiments, inspiring hopefuls like a young Michael Faraday to consider a career in science. In turn, Faraday's Christmas lectures for kids enticed yet another generation of enthusiastic minds to catch bugs, play with magnets and mix together the odd selection of dangerous chemicals in their shed or basement.

The Royal Institution's intentions were fundamentally practical— to improve Great Britain's resource production through applying scientifically tested ideas. As such, science became respected as a tool of innovation rather than just a way of making sense of nature for the love of knowledge. "Why, sir, there is every probability that you will soon be able to tax it," Michael Faraday once said rather facetiously to the Chancellor of the Exchequer regarding his discoveries about electricity. In attempting to convince the British public of the virtues of studying nature, science for the sake of pure discovery came a distant second to using it to improve technology.

Lectures were exciting affairs full of explosions, kaleidoscopic solutions, colored smoke, steam and sparks. In 1838, a similar institution called the Royal Polytechnic opened in London, where, as later claimed in its prospectus, "the public, at little expense, may acquire practical knowledge of the various arts and branches of science connected with manufacturers, mining operations and rural economy." The Polytechnic was a mix of science museum, working laboratory and lecture hall. Major attractions included rides on its working diving bell and demonstrations of early photography.

The industrial revolution made it difficult for the public to distinguish technological equipment from the very process that produced it. Few citizens cared about the contention over experimental data, questionable conclusions or vague hypotheses speculating on the elusive nature of reality. All that mattered was the solid end product—stronger engines, faster printing presses, harder metals, hardier materials, healthier crops, brighter dyes.

People all over the westernized world were falling in love with what science could do for them. Colonization had brought back exotic goods and animals from far-off lands, while the gentleman's pastime of rock collecting had produced the exciting new field of paleontology, hinting at the existence of giant lizards that once walked the Earth. Dinner parties buzzed with talk of Gideon Mantell's giant reptile bones, man-apes from the Congo, electrical cures and drugs that could dissolve pain.

Science had become society's latest form of entertainment.

FOR THE LOVE OF SCIENCE

"On one of the planets that orbits the star named Sirius," the 18th-century satirist Voltaire wrote, "there lived a spirited young man, whom I had the honor of meeting on the last voyage he made to our little ant hill. He was called Micromegas, a fitting name for anyone so great. He was eight leagues tall, or 24,000 geometric paces of five feet each."

As old as the story is, *Micromegas* wasn't the first to involve an account of a voyage through space. In fact, as far back as the 2nd century, the Syrian author Lucian of Samosata wrote of a band of travelers who were blown onto the moon by a waterspout, where they encountered strange beings such as centaurs and dog-faced men on flying acorns.

Both stories would easily pass for science fiction in any modern bookshop, yet both were more akin to satirical commentaries than tales inspired by scientific discoveries. Lucian was making a statement on the absurdity of representing mythology as truth, while Voltaire's piece relied on the perspective of an entity from a dramatically foreign culture to give voice to his comments on western society. The space themes were little more than convenient devices.

Of course, it all depends on how we define science fiction. The author Robert Heinlein suggested science fiction constituted "stories that would cease to exist if elements of science or technology were omitted," which is also reflected by popular writer Isaac Asimov's opinion that science fiction stories "feature authentic scientific knowledge and depend on it for plot development and plot resolution." The honor of the world's first true science fiction story could therefore arguably belong to the astronomer Johannes Kepler for his early 17th-century short story *Somnium* ("Dream"). In it he describes the Earth's movement as if seen from the lunar surface. While it may lack the charaterization of Voltaire's alien or the gripping plot of Lucian's narrative, *Somnium* is considered by the likes of Asimov and astronomer Carl Sagan to be the first example of a speculative work of fiction based firmly within a scientifically derived idea.

As an actual genre of literature, science fiction wouldn't become popular for another few centuries. Authors would continue to propose fantastical journeys to the moon or other lands, or push the limits of new technology in their tales, yet stories that attempted to resolve the question "what would happen if we learned that . . .?" would be a rare find until the latter half of the 19th century.

Mary Shelley's masterpiece *Frankenstein* helped to solidly establish a style of writing initially described as science romance. First published in 1818, the story is alleged to have been inspired in some part by scientists' endeavors to identify that which distinguished life from non-life. With apparent progress having been made in the previous century by the likes of Luigi Galvani, many anatomists felt that electricity was somehow responsible. The themes of Frankenstein might not have explored the fundamentals of biology in any great detail, yet the book's personal portrayal of a physician's success in reigniting the extinguished spark of life established a trend in writing which focused on the trials and tribulations of the scientific endeavor in a context of discovery.

Today, science fiction is undeniably an immensely popular literary genre. According to a 2001 survey by the National Science Foundation in the United States, nearly a third of both American men and women claimed to read books or magazines of a science fiction genre—with one in six admitting to reading the genre regularly. Science has become a solid part of our storytelling repertoire, mirroring themes common in our mythological heritage. Hollywood churns out dozens of movies every year that speculate on the future: postulating the result of space travel, encountering aliens, and dealing with robots, deadly plagues or a range of scenarios that "feel" scientific in some way.

Oddly, there are fewer stories produced that feature the science of ages past. We enjoy speculation more than the actual discovery, it seems.

Of course, the science fiction author Arthur C Clarke's third law comes to mind here. In his revised collection of essays *Profiles of the Future*, Clarke suggests that any technology sufficiently advanced is indistinguishable from magic. Key differences between sorcery and science become almost trivial as the scope of our knowledge expands beyond the horizon. All that distinguishes technology from something more miraculous is that technology simply feels like science, while magic feels like mythology. Crooked wands made from elm and decorated with crystals and feathers clearly belong in a fantasy novel. Yet a glowing metallic tube that can perform the same spectacular feats in the hands of an alien time lord is suddenly taken to be science fiction.

Likewise, the practitioners of science also reflect the role of wizard or magician in our modern and future world. Although science might be a trustworthy and respectful way of investigating how our universe works, scientists aren't always seen to be like "normal" people. They are at best portrayed as aloof, out of touch with the community and too stoic for their own good. Like the tribal shaman of many pre-industrial cultures in the northern hemisphere, the scientist is at once imbued with special knowledge and cursed to remain on society's fringe. However, unlike the shaman, they are unconnected with the universe's spiritual elements. They are frequently described in popular fiction as heartless, bent on discovering universal secrets that are best left

The Pitiful Monster   81

hidden, stopping at nothing to attain their goals. At worst they are insane and reckless, completely without ethics or compassion. Clarke's first law describes the relationship between the public and the stereo-typical, fictitious scientist as one of earnest optimism; the esteemed scientist is regarded as right when stating something is possible, but as almost certainly wrong when they claim something is impossible.

For centuries, alchemists have been associated with creation, often accused of acting in mockery of a god. The alchemist and occultist Paracelsus boasted that he could create a tiny person or "homunculus" using little more than horse manure and various bits of human tissue. Calculus in the time of Queen Elizabeth's astrologer and necromancer, Dr John Dee, was closer associated with divination than with mathematical equations. Scientists are commonly associated with pushing the boundaries of what is considered by others to be ethical, daring to presume an air of arrogance in their quest for knowledge.

In 1908, the American physiologist Charles Guthrie claimed to have removed the head of a juvenile dog and transplanted it onto an adult dog's body at the neck. While the procedure's duration reportedly damaged the young dog's brain, the animal was alleged to have still responded to stimuli, demonstrating it was possible to keep a living thing alive artificially while its heart and lungs were no longer contributing to the health of its central nervous system. Such experiments would incite outrage today, because animals are generally afforded greater protection against potentially frivolous abuse masked as research, regardless of the benefits they might afford human health.

The ethics of scientific research have always inspired great debate. Our sentiments toward specific issues often change with time—few people would feel the same disgust at accepting a donor's organ today as many would have in the middle of the 20th century. In-vitro fertilization provoked a storm of controversy when it was first made available in the 1970s, in stark contrast to its relative acceptance in the modern age. Stem cell research and genetic engineering are contemporary topics that polarize opinions in the community, yet it is possible that in decades to come, few will give as much thought to whether their study is immoral.

As such, science is often confused for the politics that governs its application, and scientists for egotistical mortals playing at being a deity at the risk of others' health and safety. Values are quickly associated with particular avenues of research, and scientists judged as saviors or destroyers in light of their methods and discoveries.

As an applied concept, science has fortunately come to be primarily associated with positive attributes. Contrasting with the image of the mad, irresponsible scientist is the fact that we often celebrate the achievements of scientists. Between 2001 and 2010, no less than half of the Australian of the Year award recipients were individuals recognized for their contributions to scientific research, specifically in fields of mental health, climate change, immunology, medicine and Indigenous health.

For the most part, our community demonstrates an appreciation of things that seem scientific in nature. A study performed by Swinburne University in 2009 found the majority of Australians believed science was continuously improving their quality of life, were mostly comfortable with the rate of progress and trusted scientific institutions and non-commercial media for information on advances in science and technology. Science appears to be more often associated with beneficial advances than with destructive or detrimental activities. A poll conducted in the United States in the late 1980s by the National Science Foundation found that over 80 percent of people expressed interest in science discoveries reported in the media, compared with only 70 percent interested in sports-related news. That figure leapt up to 90 percent if the science was specifically medical in nature.

What interested the researchers involved in this particular study was the percentage of people—43.3 percent, to be precise—who confessed to having no confidence in their understanding of science. By comparison, less than 30 percent of those who admitted to having an interest in sporting news expressed low confidence in their knowledge on sporting affairs. Given that an earlier poll conducted by the National Science Board in 1988 indicated less than half of Americans understood the Earth revolved around the Sun once every year, and even fewer were aware that antibiotics were useless against viruses, it might not be all that surprising. As a community, most people typically like science as a topic—they just don't really feel confident that they "get" it.

Is this a problem? Why should it matter if the community is divided into those who understand science and those who merely appreciate it? In 2010, the Australian federal government proposed a strategy for improving the public's engagement with science. The relevant report claims that a higher level of public literacy in science promotes innovation and inspires individuals to realize the benefits that scientific investment brings to the economy, society and the environment. Scientific literacy is presented as a pragmatic issue of furthering technological progress as well as an ethical issue of people having a right to know how their tax money is being spent in research. It might also be argued that the public has a responsibility to question the values and ethics upon which science is performed, regardless of who foots the bill. On a community level, at least, it can be considered important for all members of society to have a relatively good understanding of the science others are doing for practical and ethical reasons.

Scientific literacy is a rather noble ideal. Achieving it, however, is problematic thanks to our tribal brains. If science is equated with knowledge, then communicating facts, figures and theories should be a way to increase the public's level of engagement with it. However, this boils down to the authority distributing the information. Who do you listen to when there are conflicting sources? Our brain's desire for certainty and its tendency to evaluate new information based on social clues means anybody painted as an expert, who sounds confident, shares our values and flatters our expectations, is more likely to win over our opinion . . . regardless of the scientific merits of their argument.

If we consider this to be a serious problem today, it is far from a novel one. Michael Faraday was scratching his head over the very same issue over 150 years ago.

TURNING THE TABLES

Among Victorian England's social elite, conversing with the dead was deemed to be a popular form of parlor entertainment to follow one's lavish dinner parties. Self-proclaimed gifted individuals would sell their services of clairvoyance to anyone who could afford it and treat them to a candlelit evening of table knocking, ghostly voices and channeling of expired personalities. Of course, the question of whether it was merely a creative performance or a truly mystical display was understated; all that mattered was that it was a jolly good bit of fun between dessert and the nightcap.

Michael Faraday failed to see it this way. As a popular scientist he was frequently on the receiving end of questions involving the physics behind these spiritual shenanigans. The biggest craze during his time as a public speaker was "table turning": the phenomenon by which an otherwise ordinary table would rotate freely during a séance.

Many assumed that Faraday's work on magnetism would offer an insight, obviously given that it, too, described a "spooky" invisible force. Yet ever the true scientist, Faraday needed to be quite certain that there was something to explain in the first place. He suspected that the table's movement resulted from unconscious pushing at the hands of the séance's participants. Therefore, he patiently watched a table turning event at a séance before setting up a simple device that would alert those present to any pushing or nudging that might be occurring. As predicted, where initially the table had rotated rather eerily, it stopped when the participants had their attention drawn to their own unconscious actions.

There were ultimately two possibilities. One was that any spirits who were capable of turning the table previously were now unable (or unwilling) to repeat the act. Or, two, the participants were now aware of their collective, subtle nudging and could refrain from contributing to the turning of the table.

Needless to say, the scientist's efforts went practically unnoticed.

The following week the editor of the Westminster Review wrote to Faraday and asked, on the ghostly topic, "Are we not on the eve of some new discovery in dynamics?" Others pointed out stories which most certainly couldn't be explained away as easily as the subconscious shoving by the fingers of the séance's circle.

Across the Atlantic, the respected American chemist Robert Hare was so intrigued by Faraday's explanation that involuntary muscular movements were responsible for table turning, he set to work devising his own equipment to test claims of spirit communication. The result was the "spiritoscope"—a wheel with letters around its circumference attached via more wheels and pulleys to a small board that could be gently tilted by a person wishing to communicate with supernatural elements. A subject, seated at the table, could not see the face of the wheel so had no idea what would be spelled out through their otherwise "involuntary" nudging. Far from debunking spiritualists, on using it to contact his deceased father Hare became a solid convert.

Unfortunately for the chemist, Hare found himself ridiculed by his peers for "adhering to a gigantic humbug" and left Harvard University the subject of scorn and derision. Yet he remained convinced by his observations, perhaps even more so since he had set out to disprove what he had once called a "gross delusion."

In a public lecture Faraday said, "I do not object to table-moving, for itself; for being once stated, it becomes a fit, though a very unpromising, subject for experiment; but I am opposed to the unwillingness of its advocates to investigate; their boldness to assert; the credulity of the lookers-on; their desire that the reserved and cautious objector should be in error; and I wish, by calling attention to these things, to make general want of mental discipline and education manifest."

Faraday's highlighting of a deficit in education was oft repeated. His response to a letter in the Times would fit perfectly in any modern editorial: "I think the system of education that could leave the mental condition of the public body in the state in which this subject [table turning] has found it must have been greatly deficient in some very important principle."

The amount of science within that period's education system was a watered-down version of what the average young teenager might experience today, consisting of little more than some mathematics and basic astronomy. Rarely was anything like chemistry or physics seen in the curriculum, let alone biology or human anatomy. Scientific philosophy was completely amiss, apart from an occasional foray into logic or rhetoric. Faraday's petitioning fell mostly on deaf ears. As a subject in its own right, science entered the British curriculum only in 1850, established by a gentleman named William Sharp, whose work became the model for the subject. In 1867, the British Academy for the Advancement of Science published a report that called for students to be taught more pure sciences, as well as promoting a "scientific habit of the mind." The United States would have to wait until the 1890s for the beginnings of a standarized science curriculum to be formed.

Faraday would have been pleased with the progress that eventually followed. Today, no secondary school curriculum would seem complete if not for a healthy dose of chemistry, physics and biology. Many adolescents can even experience specific disciplines such as psychology or Earth science. Most major cities around the world have museums and science centers. Television, radio and the Internet are overflowing with nature documentaries, flashy demonstrations of popping chemicals and sparking wires, and wacky "mythbusting" programs. Science entertainment is everywhere.

Some would argue we still have some way to go; science journalism often takes a back seat to sports and politics, in strong contrast to the expressed interests of the public. In 2008, America's Cable News Network (CNN) cut its staff of science, technology and environment experts in order to absorb the unit into a single editorial team. Few media groups have distinct science units as they might have for other disciplines such as business or foreign affairs. Yet while there is always room for improvement, the very fact such actions have been openly criticized and discussed demonstrates how deeply science has embedded itself in our modern culture.

If Faraday hoped the proliferation of science in schools and an increase in the public's awareness of scientific topics would see the end of table turning, he would be sadly mistaken. It's not hard to find ghost hunters postulating the electromagnetic properties or quantum flibbertigibbets of ghoulish haunts, media personalities who claim to be able to contact grandma from beyond the grave or believers in all forms of psychic powers. It takes a lot more than a basic understanding of a few scientific facts to dissuade us from taking such an active interest in the possibility of spooks, spirits and séances.

THE BEAUTICIAN'S LAB COAT

Most upmarket department stores have them—an archipelago of glass counters, which compete to sell their brand of beauty product. Invisible clouds of scent make you realize what it's like to be an insect surrounded by shouting matches of pheromones. Fortunately for me, men are usually ignored in favor of our fairer companions. The painted attendants focus their attention on passing female customers, suggesting free samples of eyeliner or waving their hands over sparkly skin foundation that will rejuvenate their wrinkles and make them look five, ten or fifteen years younger.

At one major store, I've noticed the representatives of a particular brand of make-up attired in plain white, thigh-length cotton coats. Hardly the most fashionable of uniforms, they look remarkably like the dustcoats scientists wear to protect their clothes from the nasty stains, heat, frog guts or whatever else it is they might be knocking about in the laboratory.

Why? Is it that they're concerned about getting smears of poppy-red lip gloss on their shirts? Perhaps, however nobody else seems all that bothered—behind the next counter stands a pair of young lasses in white ruffled blouses and tan skirts that are begging for a stray dusting of rouge. It's a safe bet that these beauticians wore laboratory coats for much the same reason as Michael Faraday was asked to give his opinion on spinning tables. If it sounds "sciencey," you can trust it.

Things that appear to be scientific in nature are associated with certain characteristics of rigor and authority. Public science education has definitely been successful in communicating one thing: science produces knowledge, and knowledge is more useful than guessing.

A person in a lab coat represents the authority we associate with specialist knowledge. Young children in western nations such as the UK and Australia have been shown to typically describe a scientist according to a defined stereotype that includes a white knee-length coat, glasses, untamed hair and (most worryingly) features of an older male.

Of course, when asked to create a representative image, that is what most of us would respond with, regardless of whether we put a lot of faith in the stereotype or not. We all readily acknowledge that scientists don't have to wear lab coats, have strange hair or be male. But the symbolic icon has established itself in our collective psyche—scientists are old men with frizzy hair decked out in white coats. And it's an image we've worked hard to create.

In June 2009, delegates at the general meeting for the American Medical Association put forward for discussion a proposal that doctors should be banned from wearing lab coats on the back of findings that it was a potential source of contamination. British doctors are subject to such dress code regulations, with ties, jewelery and lab coats prohibited. Wearing the same garment repeatedly between consultations with various sick people seems to be an effective way of spreading some types of pathogen. So why did physicians wear them in the first place? Was it to keep patients' blood off their tie? If so, why make them white, instead of black or blood red?

Medical professionals began to wear white coats at the turn of the 20th century, adopting the practice from researchers who had long worn simple beige coats in the laboratory. The reason seems to be quite straightforward; it gave physicians an air of respect and symbolized a change in medical culture. Through public lectures, changes to education and obvious signs of improvement in technology and medicine, people were gaining trust in the reliability of science to produce useful knowledge.

The final decades of the 19th century saw medicine become a focus for scientific discovery. In preceding centuries, the physician adhered dogmatically to the authority of historical figures. Innovators such as the rebellious 18th-century Scottish surgeon John Hunter might have dared to challenge presumed truths and criticize established beliefs, yet such audacity was rare in an era where bloodletting and enemas cured nearly everything. For the most part, medical practice was informed by ancient assumptions and deference to a rigid pecking order.

Major discoveries in microbiology and epidemiology led to improved healing and hygiene while surgeons began to benefit from novel anesthetics, detailed anatomical dissections and improvements in medication. It was a war between tradition and innovation, and science was gradually winning out over the established dogma. Physicians took to wearing the identifiable lab coats as a form of scientific uniform. While tan, green or beige were colors typically used by technicians in the laboratories, black was the choice of color for physicians doing rounds through the wards. Given hospitals were institutions for the very poor and very sick, it comes as little surprise that those who worked with dying patients adopted clothing more suitable for a funeral. The change to white lab coats was purely a revolution against this old way of thinking—a way of choosing life over death while riding the coat-tails of scientific credibility.

Just as with our physicians, the women at the beautician's counter were implying that there was more to their services than guesswork and make-believe, thanks to the stereotype that says scientists do more than just make stuff up.

Of course, the beauticians could do a lot more to illuminate their products and services in a scientific light. It all comes down to choosing the right words to describe them.

> Men and women in every branch of medicine—113 597 in all—were queried in this nationwide study of cigarette preference. Three leading research organizations made the survey. The gist of the survey was—what cigarette do you smoke, Doctor? The brand named most was Camel!

Convinced? Camel cigarettes made good use of this purported survey. It featured in a long run of advertisements, like this one from a magazine in 1946, touting its impressive results. Few doctors today would be so supportive of such research, but half a century ago cigarette companies were hardly bothered by the damaging information that was starting to link smoking with a range of health problems, from emphysema to lung cancer. Making their advertisement sound scientific gave it credibility.

While we might not find cigarettes being advertised quite like this anymore, it's not difficult to run across the same sales tactics elsewhere. Marketing groups love to include surveys, statistics and scientific facts in their campaigns to entice the consumer into purchasing their product or service. Sex, it seems, isn't the only thing that sells. Science gives a pretty face and long legs a run for their money any day.

The trust we place in information that carries a connotation of rigorous analysis is rarely conscious. Few consumers would go out of their way to buy a product purely on the back of a survey that says it's better than the rest. But in a competition for attention, even a subtle hint of scientific support can be enough to bias the most discriminating mind, even if that support is essentially meaningless.

Nobody says Camel didn't commission such a survey, or that the results weren't in their favor. Advertising laws dictate that companies cannot tell bold-faced lies. Before you think that means you can believe everything you read, however, it does leave a lot of wiggle room for them to paint a rather misleading picture.

Over 100,000 doctors certainly amounts to a big survey. That must make it significant, right? Perhaps, but it amounts to nothing more than a red herring. The importance isn't necessarily in the size of such numbers. Imagine if there were 500,000 doctors in the total global population who could smoke Camel cigarettes if they wanted to. Now, imagine if those surveyed weren't just random doctors, but were selected; maybe packs of Camel cigarettes purchased at the tobacconist came with the survey attached, for example. This would make it harder for doctors who smoked another brand to even respond in the first place. In which case, a fairer survey of fewer people would represent the figures more accurately.

Of course, this might not be the case at all with the Camel advertising campaign. Not knowing anything about the survey, I can't comment on whether it's accurate or not. The "three leading research organizations" might have been very careful with how they selected their samples, making sure they covered a wide variety of doctors. The point, however, is that it's impossible to tell simply from the advertisement alone.

Scientists publish their findings as reports in journals for a good reason. By showing how they arrived at their conclusions, readers can determine for themselves how confident they should be in the study's results. Snappy slogans with percentages and appealing facts tell you absolutely nothing about how those details were determined. "Nine out of ten dentists prefer Whitey's Chalk Paste" might initially sound impressive, but without knowing anything about how the company arrived at this number, it's useless.

Similarly, companies will occasionally fund innocuous research into a product purely to provide the public with information that appears to be scientifically relevant. Take, for instance, the formula for perfectly cooked bacon, as displayed in the opening paragraph of a 2007 *New York Times* article titled "The Perfect Bacon Sandwich Decoded: Crisp and Crunchy": $N = C + \{f_b(c_m) \cdot f_b(t_c)\} + f_b(T_s) + f_c \cdot t_a$.

Far from being something a chef might readily employ for their breakfast menu, this rather perplexing arrangement of algebra and brackets has just one purpose—to look scientific. The Leeds University researchers responsible for the study could just as easily have said that the majority of people like not just the taste and smell of bacon, but a good level of crispiness and firm texture. The quantification provided by a formula offers a useless level of precise detail, much like claiming Mt Everest is 8848.0392 meters high. But for public appeal a confusing formula looks more scientific than a simple sentence. Given the research was funded by a British subsidiary of a Danish producer of pork products, it's easy to see it as a promotional exercise rather than an attempt to better understand the cultural practices of fried bacon.

Another tactic, which relies on missing information, is to emphasize a ubiquitous fact that can be applied to all similar products. Imagine how impressive it sounds to find a bottle of cyanide-free shampoo! Impressive, that is, until you ask how many other brands are also free of cyanide (but simply don't sing jingles about it). By a company's proclaiming as a fact that their product is free of something, or contains a particular desirable ingredient, there is the implication that others are somehow different.

Fear is arguably one of the most useful tools a marketing company can employ in their campaign to prompt you to buy. If companies can encourage you as a consumer to reassess your perception of a risk, there is a chance that they'll overcome your reservations so that you buy whatever it is they're selling. Making the risk sound scientific makes it seem even more plausible. Nowhere is this truer than when it comes to health products. Not only are misleading statistics and facts delivered with a straight face, but great liberty is taken with the definitions of words.

IN THE BEGINNING, THERE WAS THE WORD . . .

Language is often treated as if meaning is solid and objective. Many debates are considered solved by pulling out the *Oxford English Dictionary* and looking up a definition. While most words have a simple dictionary or "denotative" meaning, there are often hidden emotional or "connotative" meanings as well, many of which vary considerably depending on the context in which the word is used.

Connotative meanings aren't always easy to include in a literal definition and often change depending on your cultural background. In Australia, to call a child "precocious" carries a negative connotation of being somewhat above their station, and a little too outwardly sure of their talents. A likely reason for this is the cultural "tall poppy" syndrome, where humility is prized over self-promotion. In the United States, however, precociousness is complimentary. The connotation is a positive one. In both cases, the literal meaning is the same—a person who has accomplished more than would be expected for their age or position.

Effective communication is an important quality in science. Ambiguous meanings can lead to confusion and misunderstanding. To the public, it might seem rather pedantic to argue about whether Pluto is a planet, planetoid, dwarf planet, or a Kuiper belt object. Yet precision in definitions reduces the risk of assuming qualities that the communicator didn't intend. While mythology can promote diverse meanings, science strives to shed them.

This often leads to the formation of jargon and technical phrases or a reliance on cold, clinical numbers rather than more qualitative terms. Unfortunately, while it's useful to have precise language, jargon can isolate people who don't understand its full meaning. Although such terms might initially be created to avoid colloquial baggage and misassumptions caused by a history of connotation, sooner or later they too will come with associated meanings that vary depending on your social background. It's an endless evolution of words, from creation to adoption to adaptation.

"Theory" is just such a word. Scientifically, it is still understood to be an idea that is useful in its ability to predict future observations. There is nothing more robust, more reliable, than a scientific theory. While it's possible for a theory to be proven completely wrong, it very rarely occurs. More often than not, theories are tweaked as we discover additional facts and learn more about their context, becoming a little more complicated as we are forced to take into account novel discoveries.

Outside of a scientific culture, conviction in the meaning of the word "theory" falls dramatically, equating it with a vague thought or an unsupported belief. It's something a person would throw out as a conjecture or an opinion. Those who wish to cast doubt on the possibility that evolution accounts for the diversity of life forms on Earth can say "it's just a theory" and have it equate with any other speculation.

Sometimes, borrowed words don't just change meaning. They lose it altogether. Toxicology is essentially the study of how chemicals interfere with the normal workings of the body. In other words, toxicologists study toxins. What, then, is a toxin?

Paracelsus was a 16th-century Austrian physician, alchemist, astrologer and dabbler in the dark arts. He is also known as the father of toxicology, and is famous for his phrase, "All things are poison and nothing is without poison, only the dose permits something not to be poisonous." Hence, for the better part of five centuries it's been well known that it's not so much that some chemicals are toxins, per se, but rather that "toxic" refers to the behavior of certain chemicals when they reach a particular concentration. The term "LD50" is used to refer specifically to a chemical's toxicity. It stands for "lethal dose 50," and refers to the dosage it would take to kill at least half of a population of affected individuals.

Step into any pharmacy or health food shop today and you'll find numerous products that promise to rid your body of "toxins." There are a wide variety of "detox" programs sold as books or by self-proclaimed health experts. Without going into the details of whether such programs have any merit, the word "toxin" is used not in any strictly medical sense, but rather for its vague connotative association with a reduction in health. There are no specific details covering the names of the chemicals which are removed or how the products reduce a particular chemical's lethal dose. Most people don't care—toxins are simply a collective term for the bad things in your body.

Your body deals with many potential toxins every minute of every day, either by changing them and then excreting the products, or accounting for their effect in low concentrations. An enzyme in your liver called alcohol dehydrogenase readily breaks down ethanol into acetylaldehyde, which is further oxidized by acetaldehyde dehydrogenase into harmless old acetic acid (the major component of vinegar), for instance. However, there is a difference between a chemical that has the potential to do damage and one that is actively harming you. A brief whiff of the ethanol in your perfume is completely harmless, while downing a liter of vodka would most likely kill you.

Even across disciplines that are traditionally viewed as scientific, subtle variations in meaning can create confusion. Fields such as sociology and anthropology are typically qualitative by nature, relying on descriptive terms to accurately communicate the specific properties of an observation rather than formulas and numerical constants. In the search for suitable language, words from other fields are commonly borrowed as metaphors. While an analogical term can be useful, it can also create unwanted confusion.

For example, is quantum gravity a social and linguistic construct? The physicist Alan Sokal argued that it was, and in 1996 Duke University Press felt his discussion was good enough to publish in their postmodern cultural studies journal Social Text. Unfortunately, Sokal had no idea about quantum gravity's role in linguistics. Nobody did, in fact. He made it up in an effort to highlight a lack of intellectual rigor among a group of magazine publishers.

"Transgressing the Boundaries: Toward a Transformative Hermeneutics of Quantum Gravity" was a paper full of invented terms and essentially meaningless statements. At its core was the argument that the effects of quantum gravity had political implications. Words that had strict definitions in mathematics and physics were applied to a non-mathematics field in an effort to support a conclusion. The fact it was published created a storm of controversy—Sokal's hoax had appeared to demonstrate something he had long suspected; some journals would publish anything so long as it sounded right and flattered the editor's expectations. If the article "felt" scientific according to its language, it was deemed by the editors as appropriate.

Knowing how a journal selects articles and experiments to publish is important when it comes to gauging the trustworthiness of the research. After all, we can't all be experts in everything. Sokal's hoax raised questions on how some academics used language to promote an agenda and make their beliefs appear to be reliable, throwing the journal's reputation (and that of postmodern cultural studies) into question.

Words like quantum and gravity have very particular definitions in physics. They might be useful as metaphors in other fields; however, if nobody is clear on the limits of the analogies used, confusion can set in. It was this confusion that Sokal took advantage of in getting a meaningless paper published.

Metaphors have an important place in exploring scientific ideas. In fact, it's difficult to avoid using them. Concepts from one field are often borrowed to explain those in another. Sometimes, however, the language takes on a life of its own, muddying the waters and making it difficult to understand precisely what it is the author is trying to say.

Most of the time, there isn't an intention to deceive. We use common language within our social groups—if a person is part of a group that understands the meaning of "toxin" in a non-scientific sense, they are not wrong, as such, just as an Australian isn't wrong to think a precocious child is a brat or a farmer isn't just stating the obvious by claiming their apples are organic. The context is important.

However, there is the potential for misunderstanding if that context is vague or misleading. In a world where we are forced to deal with so many different people from so many different backgrounds, confusion over the context can have a variety of effects. At best, it might be the cause of slight embarrassment in an occasional disagreement. At worst, however, this confusion can inflate your sense of confidence in a bad idea, which you might use to make some poorly informed decisions.

THE PRICE OF A GOOD IDEA

It's easy to understand why we are all capable of accepting disproven theories as if they tell us something about the universe. Pragmatic ideas take time to be revealed. They often require the letting go of comforting possibilities, not to mention the risk of isolation from friends and family who mistake your contrary position for arrogance or personal judgment. The benefits of science come at a cost.

For every drug purchased from the pharmacy or administered by a medical provider, there are thousands of potentially useful medicines that never make it. Every medication once started as a niggling thought in the back of a scientist's head. As with the chance discovery of penicillin's effect on bacteria, it might have started with an observation. Maybe an existing drug that could be tweaked to do a new job inspired the idea, or a completely novel molecule modeled from scratch on a computer. In each case the thought is born with full potential of reducing a person's suffering.

Unfortunately, every new possibility is met with a world of

unknowns that need to be dealt with before anybody can have confidence that it will do more good than harm.

These ideas first enter a phase called "preclinical testing," where a chemical engineer introduces the chemical to a solution containing growing cells or a living animal such as a rat or a mouse and watches what happens. On average, preclinical testing can take three to four years to complete. Every now and then it might be fast-tracked, should the drug be considered to be particularly important.

Of every thousand ideas that show promise, only one will make it through to the next stage. The rest are eliminated because they fail to perform as expected or demonstrate side effects that outweigh the drug's potential benefits.

In the United States, an application is made to the Food and Drug Administration (FDA) before permission is granted for further testing, while in Australia an authority called the Therapeutic Goods Administration (TGA) manages the same process. In each case, preclinical data is assessed to determine the ethics and risk of continuing with tests on people.

The clinical trial itself has three phases. The first lasts about a year and involves anywhere between 20 and 80 healthy volunteers in assessment of the safety of the drug and its recommended dosages. Free from the effects of illnesses that might present symptoms, any common side effects these people exhibit during the trial can be associated with the drug. Measurements are taken on how long the drug remains in the body, what components it metabolizes into and how it is eventually excreted.

Phase two involves people who suffer from the condition the drug is expected to treat, where anywhere between 100 and 300 volunteers are monitored on the medication over the next two years. The final phase can involve up to several thousand patients, with their physicians closely watching the effects of the drug over the course of about three years. In some cases a fourth phase exists, where a pharmaceutical continues to be observed following TGA or FDA approval in order to gather more detailed information.

Only one in five clinically tested chemicals makes it to the doctor's list of pharmaceutical options when it comes to treating a patient. Recalling that only one in a thousand ideas even makes it to clinical testing on people, it means that of every five thousand inspired brain-waves a scientist has, a single one will prove useful.

The rest will be too dangerous, too expensive or too useless to be of any good to anybody.

Pharmaceutical companies therefore have to fund a great deal of testing using the few successes that make it through to your medicine cabinet. Estimates vary enormously, from as little as a few hundred million to over a billion dollars spent on trials per drug.

In addition, clinical trials themselves aren't without risk to human lives. Although they are rare, severe reactions to new drugs in individuals do occur. In 2006, massive organ failure in four male volunteers participating in a phase-one trial for an anti-inflammatory drug made world headlines, highlighting the challenges in drug testing and leading to questions about protocol and safety. In 2007, the death of a 36-year-old woman in the United States from organ failure was linked with her participation in a clinical trial for a new gene therapy to treat rheumatoid arthritis.

Are the costs worth it? The very fact such unfortunate cases make the news demonstrates their rarity. Without such rigorous testing, deaths and debilitating side effects could be far more common. In the very least, money would be wasted on medication that did nothing, providing the patient with unnecessary stress and false hopes. Without any pharmaceuticals at all, the quality of health and wellbeing for most people would be compromised at some point in their life. Infant mortality would be much higher and fewer people would live long enough to meet their grandchildren.

Science does come at a great cost. Not only do experiments demand resources, the resulting knowledge can be a powerful tool that leads to suffering for some as much as it might lead to salvation for others. Yet it empowers people with the ability to make decisions that have the best chance of producing the results they set out to achieve.

But what of bad ideas? What harm do we risk if others in our community entertain the possibility that prayer can cure cancer? Who cares if your neighbor thinks their house is haunted or your boss is convinced that they had a premonition of their neighbor's lottery win? Does it really matter if your sister has a lucky charm she takes to exams?

The price we pay for bad ideas varies as much as good ideas. We usually accept it as a part of the overall price we pay for freedom of belief and social harmony. On the other hand, when misplaced confidence in an idea gambles with our quality of life, detracts from our trust in the scientific process or enables others in our community to exploit the human condition, those costs can amount to more than anybody should be willing to pay.

BAD BLOOD

Pneumocystis jirovecii looks like yeast under the microscope. Although this fungus's tiny, round white cells are right at home in the average person's lungs, kept in check by the normal functioning of a healthy body, it's not uncommon for the fungus to turn nasty when the immune system falters. Between October 1980 and May 1981, lung biopsies taken from five men in three different Los Angeles hospitals came up positive for P. jirovecii. None of them were on drugs that might impact on their immune system, nor did they present any sign of cancer. What they did have was a virus that would gradually come to be seen as one of the most notorious diseases on the planet.

Without yet a name for the condition itself, the US Center for Disease Control initially used the names of other diseases commonly associated with it, such as "Kaposi's Sarcoma." Other acronyms, such as "GRID" (Gay-related Immune Deficiency) and "4H" (for homosexuals, heroin addicts, Haitians and haemophiliacs), were short-lived and prejudiced against the stereotypes associated with the illness. These names finally gave way by the end of 1982 to the term we currently use: Acquired Immunodeficiency Syndrome, or "AIDS."

Today there is no significant debate among immunologists on what causes this disease. There hasn't been much of an argument since a panel formed by the US National Academy of Sciences declared in 1988, "the committee believes that the evidence that HIV causes AIDS is scientifically conclusive."

AIDS is a condition caused by a microbe called Human Immunodeficiency Virus (HIV). The virus enters the blood cells responsible for finding and destroying disease-causing microbes within a person's body and uses its cellular machinery to replicate, often destroying the blood cell in the process. Virologists who study the disease are confident in their understanding of how the virus infects and destroys white blood cells and why it is such an elusive pathogen to both cure and vaccinate against. Epidemiologists are typically well aware of the connection between changes in lifestyle and the effect this has on the spread of the disease through a population. As far as scientists who study AIDS are concerned, HIV is the root cause.

Ever since the disease came to our attention, there have been people who have expressed doubt about the link. Most have gradually come to be persuaded by the mounting evidence. A minority, however, continue to maintain that AIDS is not caused by HIV at all, with some going so far as to even deny it is caused by any form of contagion whatsoever. The result is a belief that AIDS is not a disease one can catch. Such activists seek to promote their perspective through the community, publicly raising the possibility that this disease is caused instead by lifestyle factors such as the effects of illicit drugs or poor nutrition, and therefore cannot be treated with antivirals or prevented through avoiding contaminated bodily fluids.

Early skepticism toward the relationship between the virus and the disease was the result of the difficulty researchers had in identifying effective treatments. The chameleon nature of the virus eluded the best efforts of immunologists in finding a weakness they could exploit to control it. However, with the development of antiretroviral therapies in the 1990s, a proportion of disbelievers found it difficult to maintain their doubt. AIDS might not be curable, but it could be suppressed.

In spite of this, there are communities of grassroots activists who persist in encouraging what they see as healthy debate on a topic that can risk lives if one finds oneself standing on the wrong side of the fence. Since their formation in 1988, an association in Western Australia called the Perth Group has actively promoted the view that HIV has never been proven to exist and, therefore, the public should seek to continue the discussion on the epidemiology of AIDS in light of this.

In 2007, the group's leader, Ms Eleni Papadopulos-Eleopulos, together with another group member, Valendar Turner, presented testimony at the Supreme Court of South Australia in support of an HIV-positive man appealing against his conviction of endangering the lives of three women with whom he had unprotected sex. Their claim was simple; AIDS results from a chemical process within the body caused by semen and not from a transmissible viral agent. The trial's judge dismissed their testimony on the grounds that neither of the representatives was qualified to present opinions on the topic, given Papadopulos-Eleopulos's background in nuclear physics and limited academic experience in epidemiology.

Debate in science is, of course, fundamental to making it work. We should all be encouraged to weigh the evidence and personally determine a conclusion that has the best chance of producing the outcomes we want. Discussion should never be silenced, but knowing how to access accurate information in a critical manner can be the difference between living and dying.

The website "www.aidstruth.org" was established to present evidence that contradicts the assertions made by those who deny any link between HIV and AIDS. Among its pages is a rather depressing collection of obituaries of individuals who died with the symptoms of AIDS and were HIV-positive, yet who also actively denied they could possibly have the disease. Every decision they made—whether in its prevention or in its treatment, in looking after both their own health and their interaction with others—was with this belief in mind. Nigerian music star Fela Anikulapo-Kuti is just one example, whose brother was ironically in charge of his country's AIDS program. In 1997, Anikulapo-Kuti died from a disease he believed was a white man's conspiracy, which he contracted—as many do in Africa—via unprotected sex with an infected partner.

Pseudoscience has resulted in skepticism over what causes the disease AIDS, but also over how HIV spreads. While traveling to Cameroon in 2009, Pope Benedict XVI echoed the sentiments of his predecessor by stating that the nation's AIDS epidemic was "a tragedy that cannot be overcome by money alone, that cannot be overcome through the distribution of condoms, which can even increase the problem." Programs across Africa work hard to encourage a reluctant public to engage in safe sex. Unfortunately, difficulty in promoting condom use in Africa has impeded successful control of HIV. While similar programs have had success in nations such as Thailand, this has occurred only due to a threshold of condom use being met within fields such as the sex industry. It could be argued that condoms simply aren't a viable means of HIV control in Africa given the numerous obstacles. Yet the words of a respected religious figure of the Pope's standing can rapidly dissolve any ground HIV-control programs might have made, based on religious sentiments and an appeal to faith-based morality rather than scientific evidence.

South Africa has suffered from the HIV epidemic more than any other country. Yet at the International AIDS Conference in Durban in 2000, its former president Thabo Mbeki questioned whether AIDS was really caused by the virus, insinuating the role poverty plays instead. His health minister, Manto Tshabalala-Msimang, suggested traditional remedies involving lemon juice and beetroots were effective treatments, diverting attention from the efficacy of antiretroviral drugs. A 2008 Harvard study concludes that 330,000 lives have been lost between 2000 and 2005 in South Africa due to the absence of a timely drug program, citing the role of the government's policies in failing to take advantage of decreasing costs in antiretroviral pharmaceuticals.

The statistics are truly mind-numbing. According to a 2008 report by UNAIDS on the global AIDS epidemic, an estimated 22.4 million adults and children were HIV-positive in sub-Saharan Africa. That includes 5.2 percent of the adult population. In that same year, 1.9 million men, women and children were infected, while a further 1.4 million died.

AIDS is the 21st century's version of leprosy, carrying the same baggage of sin and depravity as the biblical skin disease, leading to stereotyping and ostracism. Still others who are diagnosed with HIV simply find it difficult to reconcile the possibility of a premature death and find it easier to turn a blind eye than take measures that just might extend their healthy years by a significant amount.

The price for relying on misleading information for health decisions can be tragic. In 2009, the New South Wales Supreme Court sentenced homeopath Thomas Sam and his wife, Manju, to a combined total of ten years in prison for the manslaughter of their nine-month-old daughter, Gloria. Her death was not the result of violence or hatred, or even lack of love. Gloria died because her parents made the decision to treat her eczema with a form of medication that cannot be demonstrated to be better than a placebo, continuing as her condition led to blood poisoning and then, eventually, to her death. Their confidence in their decision was unfounded, confused by a trust that was not matched by reason.

These are, of course, extreme cases. Not all medical decisions based on bad ideas lead to death or even poor health. Purchasing a bottle of homeopathic tincture for the winter sniffles is as good as doing nothing at all; by its very nature it is unlikely to have any impact on your body whatsoever. Fortunately, a head cold is likely to ease in several days anyway. Yet the placebo still has a cost that fails to meet an expected benefit.

Such costs aren't always easily apparent. We rely not just on our own experience but on those around us. Much of it seems scientific, especially if it is linked with people we presume to be scientists. Unfortunately, this distinction can often be misleading, especially when scientists themselves appear to spend so much time disagreeing on what constitutes a good theory.

TRUST ME—I'M A SCIENTIST

If you had to segregate members of the general public from the scientific community, how would you make the distinction? What characteristics would you use to distinguish one tribe from the other? What makes them so different, if anything at all?

Attending a science lecture, it might be easy to face the front and recognize the one giving the lecture as the expert. You could point out the fact that they make a living by working in a laboratory, mixing chemicals or running a voltage through a coil of wire while taking notes. They speak to others who do the same thing and discuss their ideas.

However, in the audience there would be others who read journals and books on science. There would be middle-aged ladies and young gentlemen with telescopes who follow the nightly movements of stars. We'd find adolescent collectors of insects, sports enthusiasts who are adept at mathematics, young apprentices relying on recent innovations and elderly people who have routinely attended the same Friday night lectures for decades. Importantly, there would also be those who liked science and were drawn to products and ideas that sounded scientific, even if they believed their understanding of it was limited. None of these people would write books or publish their findings in special journals, but does that necessarily exclude them from the scientific community itself?

The convenience of drawing a line between the institutions called "science" and "the public" is based on a misconception. Although there are certainly conferences and gatherings of professional researchers, clinicians, doctors and academics, there is no single, all-encompassing organization of scientists who gather regularly to vote on whether an idea should be declared scientifically true or dismissed as nonsense. The practicing of science—regardless of its definition—is not limited to one type of individual or a single profession. While there are definitely people we can all agree are scientists, they are also members of the general public. They could well shop at the local supermarket, vote for their favorite singer on *Australian Idol*, walk their dog in the park, play computer games and read the newspaper. Likewise, there are members of the public whom few would identify as scientists at all, who nonetheless share the same passion, possess comparable knowledge or contribute to research on some level.

Even among those who might put the word "scientist" on their entry visa, there are differences in values and beliefs. Tribes exist within the scientific community. Some would happily agree with everything in this book. Others will have already thrown it across the room or fed it to their dog as a chew toy. There is no more an easily defined community of scientists as there is a single community of artists, teachers or farmers.

This is a most fortunate thing. It's from this diversity that we get progress. Science moves on *because* of the arguments, the backstabbing, the complaints and the petty rivalry between different tribes, cliques and coteries within different parts of our community. It's difficult to take an idea and have it accepted immediately as a theory. From this mess of bickering and heated debates only the strongest ideas survive. While an idea might be loved because the person who imagined it is celebrated, there will always be somebody who will seek to attack it, either out of jealousy, revenge, anger or simply because the idea conflicts with what they presume to be true.

The philosopher Thomas Kuhn referred to the gradual acceptance of an idea as a "paradigm shift." As one generation disappears, another with fewer social ties to the last comes in and, just possibly, sees merit in the ideas they dismissed. Observations are reviewed without the influence of the last generation's tribes weighing them down, and those ideas that are still liked are increasingly accepted as they are found to be useful.

Arguments fade, rivalries are resolved, and only useful theories remain.

If only it were always that simple.

There will always be exceptions to this rule. Our social brains will cling to ideas for myriad reasons, in spite of the observations that might contradict them. We will lie to ourselves, cheat, plead and close our eyes. And if you think your own brain is any different, that you're exempt from such biased thinking and are incapable of being fooled by your neurological quirks, think again.

Chapter 4

THE LOGICAL ALIEN

Why are we so unreasonable?

In the fictional universe of the science fiction series *Star Trek*, Vulcans are a race of pointy-eared aliens typified by their love of logic. The character of Mr Spock demonstrates this obsession by not only speaking with an air of elitist precision and intellect, but with all the emotion of a piece of pocket lint. Reason, it seems, is the oil to passion's water . . . never the two shall mix. And yet while emotion is what fundamentally drives us to act, it is with measured thought that we attempt to predict the outcomes of our decisions. Logic is a keen instrument in the philosopher's toolbox, identifying incongruities between potentially useful ideas, yet it does not always come easily to our tribal minds.

WATER AND OIL

Like the words "reason" and "rationality," logic is associated with a healthy, functioning mind. Nobody would joyfully confess to being irrational, illogical and unreasonable when it comes to making an important decision. Logic goes hand in hand with an enlightened society and progressive culture. It's hard to picture a Neanderthal being logical. But is this view warranted? Is logic a relatively modern invention or a direct consequence of our ability to think symbolically?

While the fundamental attributes behind our ability to think logically stem from neurological functions, as a system of rules, definitions and values we can find its origins in a range of ancient cultures. China's 4th-century BCE philosopher Zhuangzi is today better known for describing a dream he had in which he was a butterfly, prompting him to question the nature of knowledge and reality. His writings represent one of the oldest examples of epistemology, or the study of how we know things to be true. An ancient collection of hymns from India called the Rigveda could date back as far as 1700 BCE, and contains a verse that was later formalized as a piece of Indian and Buddhist philosophy referred to as the *Catuskoti* ("the indivisible four," or a "*tetralemma*"), which attempts to categorize things around us as either an affirmative, a negative, both or neither.

In the western world, Aristotle is regarded as the first logician. While logic as a concept existed before his time, his system of deductive syllogism came to dominate reasoned thinking across Europe and the Mediterranean for over a millennium. Yet his system of traditional logic wasn't the only form of reason to have developed. A group of 3rd-century BCE Greek philosophers known as the Stoics had developed a system of logic based on a more formulaic method of reasoning, which came to be described as "propositional calculus." By the 19th century, Aristotle's form of traditional logic was found to be problematic, and replaced by a more precise array of symbols and rules that better reflects what we see as mathematics today.

Yet logic's foundations are as true today as they were then; whether it's deductive syllogism or propositional calculus, logic is a system of strict rules that don't vary according to how you might feel on any given day. What is logical today will still be so tomorrow, regardless of a change in culture or personal sentiments. Two people might quibble over the use of particular words and can certainly disagree over the premises for an argument, but logic has a mathematical quality that goes beyond mere opinion. Conversely, as anybody who has ever suffered unrequited love, bickered with a friend or felt crushed at a hero's fall from grace knows all too well, emotions are never stable and vary with experience and personality.

Logic is based on a relatively simple feature of our neurology—our brains innately recognize the fact that mutually exclusive beliefs cannot be equally true. I cannot be both a man and the planet Jupiter. Three cannot also be four. It can't be midnight at noon. I cannot be both dead and alive at the same time. With some metaphorical word play it might be possible to make such statements true to overcome that sense of conflict known as "cognitive dissonance," yet logic deals with an agreement on strict definitions rather than poetic language.

Premises can be formed any number of ways. Some are synthesized logically, produced as a conclusion from other logical deductions. Usually, they are self-evident or "axiomatic," much like saying all circles have a circumference described by a consistent radius, which is true because that's what defines something as a circle.

Premises can also be induced. In logic, induction is the increase in confidence we feel the more often something happens. If the postal service comes by at two in the afternoon on Monday, Tuesday and Wednesday, you'd probably check the mailbox after that time on Thursday and Friday. You've logically induced the fact that mail is delivered at a certain time each day. It doesn't make the fact universally true—there might be no mail service on the weekend, for example. But it is logical to guess it will probably happen again based on the frequency of your past experience until you have a reason to change your mind.

Deductive logic is what we do with those guesses. If we believe that mail is delivered every day, we can propose a rule and use that rule to make predictions.

Premise 1: Mail is delivered at two every afternoon.
Premise 2: It is now two in the afternoon.
Prediction: Therefore, the mail must have been delivered

This conclusion is perfectly logical. Our rule can be considered accurate for as long as there is mail coming at two each afternoon. Once we see a change that causes us to rethink our premises, we must modify our deduction.

As simple as this might seem, however, logic isn't always as intuitive. More importantly, our brain has ways of dealing with cognitive dissonance without using logic. It's easy to make assumptions using thinking that resembles logic but fails to produce consistent or useful conclusions.

For example, contrary to what it might seem, we cannot use our rule to make the following prediction:

Premise 1: Mail is delivered at two every afternoon.
Premise 2: Mail has been delivered.
Prediction: Therefore, it must be at least two in the afternoon.

It might be a fair guess based on other observations, but our initial logical deduction says nothing about whether mail is also delivered at other times of the day. False logic comes a little too easily to our brains.

Here's another example. Consider this:

Arthur is looking at Sam, but Sam is looking at Judy. Arthur is a man, and Judy is a woman. Nobody knows Sam's gender. Is a man looking at a woman?

A) Yes
B) No
C) There isn't enough information for me to determine an answer.

If you answered C, you're in good company. Not only did I come to this conclusion the first time I came across a similar problem, 80 percent of the population have as well. The correct answer is A.

We both answered "C" because we fell back on thinking patterns that usually prove to be rather good at solving problems. We used a process our brains had prepared for just such an occasion, believing it could solve this problem. At a glance, there doesn't seem to be enough information to produce an answer so we saw little point in thinking more than necessary.

Now, take a moment to consider both possible scenarios. If Sam is a man, then he is looking at Judy. The answer is yes. If Sam is a woman, then the answer is *still* yes, because Arthur is a man and he happens to be looking at her. There is no need to know whether Sam is a man or a woman for you to answer, so there is plenty of information to give the answer A.

In effect, I was trying to save energy from thinking harder than necessary and gambling on being correct. If C hadn't been offered as a possible response, I wouldn't have had any choice but to put more effort into thinking the situation through properly.

Humans, of course, aren't Vulcans. Just as Spock found it a struggle to be perpetually logical about everything, humans find it taxing to apply reason to every single decision in the face of an emotional pull. Emotions not only drive our behavior and motivate our desire to act, they can impede our sense of rational judgment. Who hasn't ever succumbed to one more slice of cake when they know they should be dieting, stolen a kiss when they know it could lead to a broken heart or gone over budget to purchase something just that little bit nicer when they know rent is due?

Determining precisely when it is permissible to let the heart win over the head is an important skill to learn. Picking songs to store on your personal MP3 player isn't always going to come down to a well-reasoned debate based on a set of robust premises—you'd do well to go with your gut and pick music that makes you feel happy. Selecting your medication in the same way, on the other hand, could have rather drastic consequences.

Every day we face situations where we need to apply reason to make a decision. Unlike a multiple-choice test in a textbook, there is no solution in the back against which you can check your answer—you're simply left with the consequences of your actions. Thinking through every single decision you make in a day could be enough to drive you insane, while not thinking some things through properly could lead to some rather dire outcomes you hadn't expected.

Let's explore a small sample of our thinking strategies, which most of us take for granted and rarely consider in much depth. While they might seem reasonable, the logic falls apart on closer inspection, making them fallacies. Ideas that initially appear rather rational can sometimes be supported more by gut feelings than logical rules, reinforced by the emotional drive to connect with others in the community. The tribal brain seeks to reinforce our bonds with others in our group in the face of broken logic, making even the most unreasonable conclusions appear to be unquestionable truths.

NATURALLY ILLOGICAL

Herbal remedies are all natural, making them a better alternative to pharmaceuticals for the treatment of asthma.

There is no better campaign manager than Mother Nature. It's difficult to avoid claims made by advertisers that their product is "all natural," "natural ingredients" or "nothing artificial," simply because, at face value, nature is a force for good while artificial is, well, not.

Historically, what philosopher George Edward Moore referred to as the "naturalistic fallacy" was any argument based on the assumption that something could be inherently and objectively positive or negative in a universal sense. For example, if an action or object was considered pleasant, productive, simple, symmetrical or expensive, it could be described as inherently "good." Likewise, things could be considered "bad" in absolute terms if they were cheap, broken, tiring, harmful or unpleasant.

Moore stated that "good" wasn't a property in the same sense as mass, color or size. It depends entirely on a context of values or qualifiers. Hedonists would love to believe that pleasant feelings made something good; however, they might have a hard time convincing their grandparents that sex, drugs and rock and roll were necessarily positive things simply because they were fun to indulge in. Each would be evaluating the behaviors using different qualifiers.

Positive and negative are subjective values we attribute to a situation, and, as such, nothing can be considered universally "good" or "bad" without supplying an additional context.

Such values can and do include moral positions, which are behaviors that are influenced by the expectations of the community. Philosophers throughout history have proposed universal systems of morality. Plato maintained that it was possible to consider something as virtuous based on an understanding of its ideal metaphysical archetype; hence, ethical behavior emerged by applying reason and recognizing the true form of something. The German philosopher Immanuel Kant objectified morality by describing a concept called the "categorical imperative" where, similar to the golden rule, a person should act in the same manner they would expect of anybody in the same situation. Other systems are inseparable from religious opinions, deferring to a supernatural entity for a system of moral laws.

Using science to not just explain the role of moral behavior, but to quantify and evaluate it, barely dates back a century or two. Charles Darwin considered altruism in animals as an evolved social trait but struggled to describe how it might benefit individuals within a group. Yet fellow biologist Thomas Huxley expressed doubt on the issue in his 1893 book *Evolution and Ethics*, claiming that the prevalence of both moral and immoral behavior makes it difficult to objectively ascribe one with greater benefits via reason alone. He writes, "But as the immoral sentiments have no less been evolved, there is so far as much natural sanction for the one as the other. The thief and the murderer follow nature just as much as the philanthropist." Similarly, the Scottish philosopher David Hume warned against confusing what "is"—as we observe it—with what "ought" to be as judged by reason.

The American author Sam Harris maintains that it is possible to objectively describe a system of moral values traditionally viewed as personal or subjective, simply by pointing out that humans share certain fundamental needs and aversions that contribute to their wellbeing. First-degree murder exacted for pure enjoyment, for example, runs contrary to the flourishing of a species.

However, even this moral argument remains contextual, as a desire for wellbeing is itself a value. "Good" could well refer to those things that make us content as individuals or as a community; however, centuries of questioning the nature of morality have not resulted in a definitive, objective system, and are unlikely to any time soon.

In an age of environmentalism and conservation, the dichotomy of nature and human activity in the post-industrial landscape has become strongly associated with moral values. Sustainability is viewed as much as a moral obligation as it is a pragmatic one, relying on sympathy for the plight of future generations.

Not only is the concept of nature wrapped up in a context of morality, the word "nature" itself suffers from something of an identity crisis. Sure, it makes most of us think of bubbling streams and rabbits and fields of wildflowers as opposed to skyscrapers and foam cups. But finding that line between natural and unnatural can be easier said than done. Take gardens, for instance; is a landscape still natural if people have taken steps to seed it with native plants? After all, in spite of being full of flowers and bees and herbs and weeds, it took human intervention to bring those elements together. Without a person, it never would have existed, therefore it's technically artificial. Yet for all purposes, it could resemble any other patch of virgin wilderness.

Human intervention of any sort might create a nice boundary if it didn't directly lead to some rather unfortunate complications. Health food products often claim to be all natural, yet those ingredients can't be found clean and ready to be mixed in the wild. It took a person to find or grow them; peel, grind or deseed them; mix them; and stick them in a container for delivery to the local supermarket or health food shop. By the same token, take any product with seemingly artificial constituents and trace those ingredients back, and you can often guarantee they all originated from some type of natural resource. At what point does a product go from being natural to unnatural?

Let's assume that the word "natural" refers less to a dichotomy and describes more of a spectrum—the more a person changes a resource, the worse it is for you. It's quite easy to find counter examples of such a belief.

Burrawang seeds are poisonous yet full of starch. They are a traditional bush food of Sydney's Indigenous Cadigal people, who wash and mash them repeatedly in water over a number of days before grinding them and turning them into cakes. This process is most certainly artificial—it involves humans modifying the original item through a rather involved process. In their "natural" state, the seeds would kill a person if consumed in relatively small quantities. Once modified, they are a safe and nutritious source of energy. Few people would consider burrawang seeds to be good for you prior to undergoing such rigorous treatment.

Traditional medicines of many Indigenous populations around the world have been used for generations to treat types of illness. Most rely on one or more active ingredients in a local plant or animal to do the trick, yet experience passed on through their culture has taught them how to get the most out of the treatment with the least risk to the individual's health. Extracting the active components or removing potentially toxic amounts of a particular chemical is often immensely useful. For instance, extracting salicylic acid from plants and modifying it with an acetyl group has provided the world with aspirin. This small change made it less irritating yet performed much the same task as the "natural" plant remedies had.

A large percentage of pharmaceutical treatments have their origins in plants or micro-organisms. Extracting the active ingredients makes it possible to calculate dosages and determine whether they might conflict with other drugs. As discussed in chapter 3, laws require that newly developed pharmaceuticals go through rigorous testing procedures before they can be deemed safe. Even so, it's possible for drugs to be made available that don't work as predicted for some individuals, or which can create unpredicted side effects. Given the number of lives saved or improved each year by one of many chemical treatments, it might be a small price to pay.

It's easy to associate food additive codes such as "food additive E260" with bad things and familiar words such as "vinegar" with good things, especially if we don't know they are both essentially acetic acid. You might not want to put chemicals like ethyl butanoate, ethyl hexanoate and 2,5-dimethyl-4-hydroxy-3(2H)-furanone into your body, unless you understood they help to give strawberries their delicious smell. The world outside the boundaries of our cities is full of things that can smell wonderful, feel good and help your body in times of illness. It is also full of poisons, diseases and things that make you feel plain awful.

Our tendency to label our surroundings according to values we've inherited from our community doesn't just save time and effort, it demonstrates who shares our moral position and who doesn't; who is of our tribe and who is an enemy. In sharing a belief that some things are intrinsically good and desirable, we better define the boundaries of our social group.

THE ILLOGICAL VOTE

*Of course it works; how could millions of
people be wrong?*

Inspired by a $5,000 reward for any information on the location of a C-46
marine transport aircraft reported missing near Mt Rainier in Washington, pilot and businessman Kenneth Arnold decided to alter his
planned route and take a look over the area where it was reported to
have vanished. It was 1947, a time when the Cold War was warming up
and the American people were sensitive to the threat of Russian attack.
The term UFO wasn't yet used to describe objects in the sky that defied
explanations and the very thought of interplanetary space travel was
yet to be taken seriously beyond the occasional science fiction novel.
So when Kenneth saw nine strange objects shimmering in the mid-
afternoon sky near the Cascade Mountains, his conclusion was
simple—it was the Russians.

On failing to get through to the Federal Bureau of Investigation
after landing in Oregon, the pilot made the decision to report to the
town's newspaper, the *East Oregonian*. This report would be historic,
to say the least.

IMPOSSIBLE! MAYBE, BUT SEEIN' IS BELIEVIN,' SAYS FLIER

Kenneth Arnold, with the fire control at Boise and who was flying in
southern Washington yesterday afternoon in search of a missing
marine plane, stopped here en route to Boise today with an unusual
story—which he doesn't expect people to believe but which he
declared to be true.

He said he sighted nine saucer-like air craft flying in formation at
3 p.m. yesterday, extremely bright—as if they were nickel plate—and
flying at an immense rate of speed. He estimated that they were at an
altitude between 9,500 and 10,000 feet and clocked them from Mt.
Rainier to Mt. Adams, arriving at the amazing speed of about 1,200
miles an hour. "It seemed impossible," he said, "but there it is—I saw
it with my own eyes."

Whatever it was that Mr Arnold saw, it's interesting to note that he was never quoted describing the objects as saucer-shaped, but rather as crescents that flew in a V formation. Where did the word "saucer" come from then? Kenneth Arnold said the objects' erratic movement could be compared with a saucer being skipped across the surface of a lake. Nevertheless, the journalist who wrote the two similar articles for both the *East Oregonian* and the *Associated Press* miscopied his notes and sparked a terminology that has lasted to this day.

Multitudes of people in the decades since have alleged that they've seen flying saucers zipping across the sky. Prior to that, references mentioning sightings of strange craft of any kind, let alone of hovering disc-shaped objects, moving across the sky were comparatively rare.

The term "Unidentified Flying Object" has since been coined to describe an array of peculiar meteorological observations, whether floating spheres, oblongs, discs, cylinders or cigars. While Mr Arnold was quick to point a finger at the Russians as the cause of his sighting, it's difficult to ignore that there is a modern connection between flying saucers and extraterrestrial spacecraft. When most people think of UFOs, they think of space travel and aliens.

Aliens of all different makes and breeds could have taken an active interest in our planet since the Cold War. That would certainly explain the increase in sightings. A more likely hypothesis, however, is that what was once ignored for lack of knowledge on the matter was now given a cause thanks to the spread of a belief.

Reports of encounters with extraterrestrial life forms always seem to match a stereotype common to the time of the sighting. Over the decades, descriptions of the alien entities have gradually evolved from the classic Nordic human in a body-hugging jumpsuit to robots, to hairy beasts, to small goblin-like creatures, to today's bulbous headed, bug-eyed "grays." While it is indeed possible that they all represent successive waves of interstellar tourists, it's more likely that beliefs are being shaped by popular perceptions.

A typical defense from those who believe in alien visitation is that so many people claim to have experienced it. Sure, some might be hoaxes and mistakes, but not all of them . . . right? Could hundreds, or even thousands, of people really be that wrong? Can so many people be delusional?

It is certainly possible. Lies, hoaxes and mistakes could indeed account for every single case. Every memory of alien abduction could most certainly be flawed. While "delusional" might be a rather strong term, everybody's brain is susceptible to making exactly the same errors.

This tendency to point out the overwhelming popularity of an idea can be found supporting anything from the success of certain medical treatments to the existence of God. On the face of it, such an argument seems valid. One person might confuse a weather balloon for something more mysterious, but surely not a hundred. A few mischievous pranksters could account for certain fuzzy photographs or stories of abduction, but what of those who seem so trustworthy? They can't all be wrong.

Such an assumption relies on knowing the probability of a particular mistake or act occurring in a population. If one person can make a mistake, there's no reason twenty can't, or a hundred, or a billion, so long as each person shares the same basic thinking processes that give rise to that error. Likewise, if enough people are inspired to gain something by fooling their peers, there is no reason that there can't be thousands of hoaxes committed over a number of decades.

The question, therefore, is this: is it more likely that alien spacecrafts explain the sightings of UFOs, or something more terrestrial? We have plenty of reasons to suspect we're all capable of making mistakes. We also have plenty of reasons to think hoaxes are common. We have nothing beyond the accounts of sightings themselves that tells us intelligent aliens with fantastic technology exist.

It's difficult to avoid using a popularity vote to support a belief, however. While it might be illogical, it sits rather comfortably with tribal thinking. Engaging with others who share our beliefs bolsters our confidence in having those beliefs. Attending church every Sunday reinforces that sense of being just one of many believers. Even atheists are finding solace in online or real world communities and conventions for want of communal support.

What seems to be widely accepted depends on your point of view. You'd know the beliefs of your friends and family more than you would your neighbors.' If your family unanimously believes in a Christian God, it's more likely you'll identify with it as a popular belief than you would with Allah or Buddha. When it comes to the opinions and views of individuals from another country or culture, it's unlikely you'd know a great deal unless you're a seasoned traveler or an experienced anthropologist. In short, the perception of a popular belief varies depending on the population you're basing it on.

There is absolutely nothing wrong with enjoying a sense of shared belief, of course. However, in and of itself, reality isn't a popularity contest.

IGNORANTLY ILLOGICAL

Nobody knows what the Loch Ness Monster is, so it must be an animal that is new to science.

At the opposite end of the popularity scale is the so-called argument from ignorance. While one is the influence of many people's shared opinion, the other is a belief that is reinforced because nobody else seems to have a clue.

UFO has to be one of today's most ironic acronyms. Literature on the topic is full of speculation on their potentially alien origins. For a flying object lacking ID, it seems that a lot of people feel they have the FO well and truly identified.

The rationale behind this bit of broken logic is simple—by definition, nobody can say with absolute certainty what causes the observation of a floating light in the sky or a mysterious photograph of a strange disc. Therefore, if no explanation can be agreed on, other, more exotic possibilities seem likely. As Sherlock Holmes was famous for saying to Dr Watson throughout his adventures, "How often have I said to you that when you have eliminated the impossible, whatever remains, however improbable, must be the truth?"

In some ways, this is true. If I give you a choice of three possible answers to a problem and two are shown to be impossible given what you know, then the third has to be the right one. Yet the universe isn't this generous. Science isn't a multiple-choice questionnaire where the right answer can be found in the midst of three or four incorrect ones, but rather one of those annoying open-ended questions you get at the back of the exam. Eliminating the impossible doesn't assure you of whittling the explanations down to a single correct answer.

Just because you cannot see any coffee in one supermarket, it doesn't necessarily mean the shop next door is now more likely to have it. A doctor who doesn't recognize your rash cannot logically come to the conclusion it must be a rare variety of catpox. If a UFO cannot be concluded to be a weather balloon, swamp gas, Venus or a Chinese lantern, it's not any more likely to be a space vessel from Andromedes. Sometimes "we don't know" is the most useful answer.

Climate change is an issue of diverse beliefs. At one absolute extreme is the environmentalist who values the eternal preservation of Earth's ecosystems in a pristine state, untouched by human hands. In their view, humans have changed the climate in ways that will increasingly lead to environmental devastation and, consequently, the extinction of the human species (if not all life as we know it). At the other end are those who contend that humans are incapable of changing something as powerful as the climate, and natural cycles are responsible for any changes we're seeing . . . including possible cooling.

In between there is a mixture of views fed by the expressions of scientists, politicians, journalists, teachers, celebrities and next-door neighbors. The average person forced to consider such a variety of opinions each day can find the debate to be a messy mix of emotion and reason.

The potential consequences of not acting to combat climate change can seem quite dire. Mass extinctions, killer weather events, the global collapse of entire economies, plagues, wars and famines all weigh on the mind. If you listen to some environmental campaigns, all of this is avoidable; simply turn off the lights and catch the bus to work in the morning. Even if you believe this accomplishes nothing, there seems to be little harm in doing this, right?

This is called "Pascal's wager." The 17th-century French mathematician Blaise Pascal revolutionized mathematics by introducing the concept of infinity to probability while attempting to demonstrate where reason falls apart in the face of discussing religion. Essentially, he argued that if you cannot be certain in the probability of an outcome, the choice becomes a dichotomy. Either God exists, or God doesn't exist. There's no middle ground (God cannot truthfully half exist).

Believing in God could mean infinite happiness (eternity in Heaven) for a finite cost (avoiding sin for one lifetime). Not believing in God could mean finite bliss (committing a few sins) for an infinite cost (eternity in Hell). The choice seems simple—you might as well believe in God for a tiny cost and be rewarded by an eternity of happiness. Likewise, using Pascal's wager to argue for turning off your lights also seems pretty straightforward. It's a small cost in the face of potentially dire consequences.

Unfortunately, neither is argument is logical.

For one thing, these arguments assume that it is strictly a choice between an action that leads to a reward and a failure to act that leads to punishment. But what if this version of "god" isn't the only deity? What if the Judeo-Christian god is the wrong deity to worship? Far from being a toss of the coin, it's actually a roll of a die with countless sides. Believing in the wrong god could lead to eternal damnation, at least as far as the logic behind the argument goes. Likewise, in the absence of additional information, thinking you've made a difference with the flick of a switch could just as easily be doing far more harm than good.

False dichotomies are often encountered in emotional discussions. Either you're with us or against us! You're either a creationist or an evolutionist! If you don't think I'm telling the truth, you must think I'm a liar! If you don't know, I must be right! In each case the situation is painted as if there are limited, known possibilities.

You can blame your left hemisphere for this need for an answer, however. As discussed earlier, the left half of your brain will go to extraordinary lengths to make you feel confident in an answer regardless of whether it is logical or not. While absolute knowledge of the future is often impossible, our actions are always absolute. You might not know for sure if a berry is poisonous or not, but actions always come down to a simple choice. You either choose to put some of the berry in your mouth, or you choose not to. In addition, hesitation for want of more information carries its own risks. In competing with others within your tribe, it's a case of the early bird gets the worm. While working with other members of your social group is beneficial, evolution continues to drive the individual in competing with other members of their community. Balancing risk of loss with the time taken to gather information can favor acting with certainty even if it's not balanced by reason.

INSULTINGLY ILLOGICAL

And you believed him?
The man beats his wife!

Prior to Hitler's ascent, Germany had made a significant contribution to scientific knowledge. Some of history's most talented physicists and chemists were of German stock and the academic tradition continued well into the establishment of the Third Reich. Physics underwent a revolution in the first decades of the 20th century thanks to the contributions of German scientists like Max Planck, Max Born and Werner Heisenberg. With the First World War, the reputation of German science became tarnished as famous civilian institutions turned their efforts to military research.

It wasn't until much later that loyal National Socialists were granted positions of authority in Germany's major research institutions and universities, and the science industry adopted the patriotic spirit of *Gleichschaltung*, or "marching in step." The effect this had on those in Germany's scientific community was unsurprisingly encouraging, even as it isolated itself from much of the world's scientific community. Nazism supported the contributions of its scientists, especially when it involved "improving" German society.

In order to achieve his academic credentials, Dr Joseph Mengele undertook a series of experiments on live human subjects. Taking Jewish twins from the concentration camps, he would inoculate them with various substances and compare their physiological reactions. He would inject chemicals into other subjects' irises in an effort to change their color. As if such inhumane experiments weren't enough, he had a reputation for executing his subjects on the slightest whim, and performing autopsies to settle arguments as easily as one might Google for an answer today.

A multitude of similarly horrific experiments were performed during the Second World War, depriving people of health and liberty in the name of scientific discovery.

There's no doubt that to many people Dr Mengele came to represent the epitome of evil. It leads to a rather interesting question—is there a relationship between the potential accuracy of a discovery and the moral repercussions of the method used to find it? Logically, it might be difficult to draw a connection between the two, yet we are often challenged to credit discovery that arose from a process we find to personally offend our moral position.

In the years prior to his dismissal from Seoul University in 2006, South Korean researcher Hwang Woo-Suk had become a respected scientist in the field of stem cell research. Articles he published in the journal Nature regarding the successful creation of human embryonic stem cells via cloning were revolutionary . . . or rather they would have been, if they hadn't been based on fabricated data. Hwang and his team of researchers basically made it up.

Unlike Mengele, Woo-Suk was found by a court to have intentionally deceived his colleagues and the public. While hardly evil, it does reflect on his trustworthiness as a scientist. On one level you couldn't be blamed for viewing all of his work in light of his demonstrated ability to engage in fraud. After all, you now have evidence that he is capable of deceit. Where your heart might encourage you to be more skeptical of what he has to say, the potential accuracy of his claims is not dependent on his truthfulness as a person. If he is wrong, it's because the claim is wrong—not because he has lied previously. Yet our view of the information and its validity is interpreted within a personal context of how we judge its author.

In determining what to believe, we often have little choice but to listen to the views of others. Unfortunately, our social brains are wired to rely on the likeability or trustworthiness of the person who is offering their opinion in order to determine correct beliefs from incorrect ones.

A 2010 study conducted by US researchers attempted to explain why the public appeared to be polarized on environmental issues to a degree that wasn't matched within the scientific community. They found individuals would rely on an evaluation of shared values and principles in order to determine whether a scientist was enough of an expert to serve as a reliable source of information. Far from being a matter of supporting or dismissing science in general, we all have a tendency to determine an authority's worth based on whether they appear to be one of our tribe.

To our ancestors, there was more value in all being wrong together than one person being correct on their own. Politicians have long understood the significance of making appearances at sporting matches, kissing babies or wearing the right color tie in order to gain support for their policies. The scientific community might not be seeking the vote of the people, but they are judged as much as any candidate in an election.

ILLOGICAL BEGGING

Theft is illegal. Taxation is a form of theft. Therefore, the government is breaking the law.

The invention of writing marks the boundary between the relative oblivion of ancient humanity and the dawn of history. The oldest surviving remnants of written language were pressed into wet clay tiles using the wedge-shaped end of a reed. "Cuneiform," from the Latin word for wedge, was a form of script developed by the Sumerians of ancient Iraq over 6,000 years ago, officially making it the world's oldest form of writing.

Of course, people had been painting and inscribing rocks, egg shells and animal bones with pictures long before then. What makes one scratching "writing" and another just a picture of some caveman's grandmother? After all, both are symbolic representations of real objects.

Writing involves a systematic arrangement of shapes in order to communicate an idea. It's the positioning of the symbols with respect to one another that conveys meaning rather than their similarity to a real world object. On the other hand, interpreting a cave painting of an ox standing next to a horse doesn't rely on your knowledge of what each symbol means—just the knowledge that something that looks vaguely cow-like is probably an ox.

Humans have a knack for arranging symbols according to sets of rules. The very fact you can read this means you understand the code I'm using. I take an idea in my head, find the right words, associate the sounds those words make with a sequence of shapes called "letters" and physically order those letters for the printer to press onto paper. You then go through the opposite process—after your eyes have converted the light from the page into nerve impulses, your brain's occipital lobe takes the inked lines and associates them with letters, the arrangement of letters is connected with sounds and the sequential sounds are associated with meanings so you then get an approximation of the idea that's in my head.

But can codes exist independently of somebody's intention? If I randomly scatter some toothpicks and see the Roman numerals XII, can I interpret it as writing the number 12? There's nothing stopping me from doing this, but it would be solely a one-sided affair. The lines "XII" wouldn't mean anything. It might look like a code, but nobody put it there with the intention of communicating a number for another's benefit.

We might think of many things as if they were codes. An animal's footprints could be thought of as meaningful symbols, making us think they were encoded with the purpose of being found and followed. A tracker could claim they "read" prints for information about the animal, and liken the shape of the toes and their distance apart to letters they interpret. However, the animal did not form them with the intention of letting a person know they were a ten-kilogram lizard heading west with a limp. Even when we can compare them to letters and words, they aren't literally codes.

DNA is sometimes referred to as a code due to the sequence of chemicals called "bases" within the nucleic acid chain. However, much like our eagle-eyed tracker might describe the impressions in the sand as code for "a goanna went this way," the word "code" for DNA is at best an analogy. It does not mean somebody wrote it.

It's not uncommon for people who believe life is too complicated to have arisen without some form of intelligent guidance to state that the DNA code is evidence in support of their belief. They might argue that

DNA is a code that stores information; since it's a code, it must have a creator.

This fallacious way of thinking is known as "begging the question" or "presuming the premise." The person making the argument takes for granted that their premises are correct before using them to deduce a conclusion. In this particular example the presumption is made that DNA is a code in the same way writing is a code. The conclusion that the person is trying to establish (codes have creators, DNA is a code, therefore, DNA must have a creator) is already assumed in the premise that DNA is a code, making the whole argument circular and therefore illogical.

Taken to an absurd level, begging the question is a little like saying, "My premise is correct; therefore, I am right." Logic is only useful in reaching a shared conclusion when two people can also agree on the starting premises. Yet communicating these premises can misappropriate connotations that don't exist, allow for assumptions that are unsupported and permit dubious facts to slip in unchallenged.

In some ways, begging the question and other forms of circular reasoning stem from problems of language, or as Francis Bacon referred to it, as the idol of the marketplace. As language often evolves by way of metaphors—borrowing symbolic meanings as new concepts arise and seek description—finding ways to relate concepts to one another logically can be confused by unintended meanings. Stating "dogs are carnivores because they eat meat" is technically a form of begging the question, even if it would normally pass as a reasonable statement to make. We might categorize them as carnivores because they have been observed to eat meat, but the reason why they are carnivorous has to do with their evolutionary history.

We take for granted our ability to communicate with our close friends and family using a common language. Logic evolved over the centuries into a system of numbers and formulas in an attempt to reduce the impact idioms, similes and connotation can have in creating a reasonable argument. While computers might have a better time in avoiding the begging of questions, the rest of us have to make do with finding the right words to construct reasonable arguments.

ILLOGICAL CORRELATIONS

Of course I'm a good scientist—my shed is full of the best equipment!

Nobody knows who John Frum is, let alone whether he ever existed. British naturalist David Attenborough was once told, "He look like you. He got white face. He tall man. He live long South America." To the traditional inhabitants of the island of Tanna in Vanuatu, however, John Frum is portrayed as a savior who will bring them wealth and prosperity.

February 15 is John Frum day. People from all over the island come to the village of Lamakara near Sulfur Bay to partake in a ceremony that resembles a mid-20th century American military display. Participants with the letters USA painted on their chests march in a procession with lengths of bamboo held to their shoulders as symbolic rifles, while the gathering crowd cheers and shouts.

One theory explaining the origins of this annual ritual suggests the exalted figure's name evolved from a chance meeting in the distant past with a westerner who introduced himself as "John from" America. During the Second World War, when Vanuatu was still under British and French rule and known to the western world as New Hebrides, the American military used a number of Pacific islands to store supplies for their campaigns against the Japanese. Airstrips were cut out of the forest for the cargo planes to deliver food and equipment. To the inhabitants, such foreign behaviors could only be related to as a ritualized appeal to the deities for goods that would later be delivered by aircraft.

These "cargo cults" can be found in a number of island cultures, and may even date back to the middle of the 19th century when colonial shipping vessels charted the islands. While the specific rituals vary, they tend to reflect the same imitation of foreign behaviors in order to regain the legendary riches that once came from the sea or sky. In some cases, not only are airstrips cleared in hope of a return of the aircraft, flagpoles are erected and fake technology is constructed in order to create a simulacrum of history.

Without a broader context, the island's inhabitants are left with simply a correlation between the appearance of strange food, technology and medicine and the peculiar behaviors of foreign intruders. The closest they can relate the foreign behavior to is their own rituals. Their mistake is a common fracturing of logic—they confuse a correlation between two events for causation. To them, it's the presence of a runway, uniformed soldiers and a command post that causes a plane to come down from the sky full of goodies.

No matter how often one observation follows another, it does not mean you can state in absolute terms that one event caused another. Science would be much simpler if proximity in time and space definitively equated causation. There are many reasons why two events that occur close together over and over again might be related without one having directly caused the other. For instance, there could be an unconsidered third factor that was directly responsible for both events. Or your attention might be drawn to an event only when it's accompanied by another. You might notice a bad smell in the air only when your dog walks in; in truth, the smell comes and goes frequently during the day, but you only think of it when your dog is present, making you think they were responsible for it.

For the most part, identifying the cause of an observation isn't important in our day-to-day lives, so long as we can spot a relationship. Every time you open the refrigerator door a light comes on. Who cares if it's an electrical circuit and not the fridge fairy? The same thing happens faithfully over and over again, meaning you can accurately predict that opening the door will turn on the light regardless of the actual cause. Until, of course, that fateful day arrives when the light doesn't come on. Is your fairy dead? Or do you need a new light bulb?

Nature isn't always neat and tidy. Thunder isn't always seen in sequence with lightning. Sometimes a flash coincides immediately with an almighty boom. Sometimes we don't see the lightning at all. Even cause–effect relationships can be complicated by a number of factors. Our pattern-seeking brains can cope quite well with this. Too well, in fact.

In a paper titled "Superstition in the Pigeon," the behavioral psychologist BF Skinner discussed how even this humble bird was capable of making simple correlation errors. A hungry pigeon was placed into a box containing a small hopper of food pellets that could be opened and closed. At irregular intervals the hopper was opened for five seconds. Skinner knew that pigeons could be trained to press a button or observe a light in order to access this hopper. However, in this situation, there was nothing the pigeon could do to influence the presentation of food.

Oddly, in six out of the eight cases he observed, the pigeon acted in an increasingly peculiar manner. One would circle in its place anti-clockwise. Another poked its head into the corner of the cage repeatedly. Two others swung their heads left and right in a "pendulum" fashion.

In a process known as "operant conditioning," the pigeons had somehow associated their performance of a random movement with the opening of the food hopper. Of course, there was no such correlation. Their strange movement did not really cause the hopper to open. Nonetheless, occasionally the two events coincided enough for the pigeon to make a connection.

Food wasn't delivered after every swing of the head, turn of the body or stretching of the neck. In fact, the pigeon could perform this behavior a number of times before the hopper finally opened and it could eat.

It's not difficult to compare this behavior with our own superstitions. Gamblers can be seen tapping the side of their poker machine a set number of times, while sports stars will often wear the same underwear or socks to important matches. In a world where we cannot see causes, we do our best to find correlations that help us to predict future events. Logically, however, just because "B follows A" it doesn't necessarily mean that "A causes B."

Our ability to match patterns benefits from our sharing of knowledge. In collecting observations as a community, we can find more situations where a pattern can produce better predictions. On the whole, the numerous eyes and ears of our tribes can make good use of numerous correlations. The cost is small, but significant—the correlations that are meaningless can often be difficult to dispel.

THE BLANK SLATE

Logical fallacies don't make information wrong. A scientist can still be technically correct, even if they're appealing to the popularity of a belief among fellow researchers. If a politician begs the question or encourages you to listen with Pascal's wager it doesn't mean they're lying or misinformed. They could very well be correct, even if their reasoning is illogical or poorly communicated.

But how would you know? Their speeches might be full of words that make them sound intelligent and their conclusions may even come across as perfectly reasonable. They could be charming and well educated. A PhD might sit next to their name. Without a logical argument, however, what "feels" right could amount to little more than hot air for all anybody truly knows.

Everybody starts from a null position at some point in their life; a blank slate devoid of opinion or belief. To begin to develop a belief, we can either look to personal experience or take another person's word on it. Usually it's a mix of both—our personal experiences influence how we interpret what other people tell us. We make use of the instruments within our own, personalized philosopher's toolbox to determine whether what we're told has merit or not, often using broken logic to evaluate our more cherished beliefs.

However, being left with nothing is an uncomfortable feeling; the left hemisphere makes it incredibly hard for us to ever say "I don't know" and leave it at that. It will seize the smallest amount of information and use it to gamble on a decision, regardless of whether it is truly useful or not.

What's more, our tribal brains make terrible lenses for information, distorting its accuracy depending on our affinity for its source. We will evaluate the usefulness of an idea based on whether the individual telling us appears to reflect our moral values, our sense of good and bad and our political goals or religious faith, regardless of our respect for science. Language will further impede our attempts to investigate the facts, distorting truths through innuendo or connotation, either intentionally or as an accidental consequence of cultural relativity.

We rely on the opinions of others to inform our own beliefs. But not all opinions are equal. Not all guesses have the same chance of success. Some guesses are far worse than no guess at all.

Chapter 5

THE CLEVER HORSE

What is this thing called "evidence"?

Alone with their books and experiments, the solitary genius is a common archetype, stemming from our perception that learning is an isolated act. While we construct knowledge using personal experience, we model our learning behavior on those around us, influenced by our community as we discriminate truth from fantasy. Knowledge and the forces that produce it are products of our social machinery, churning out an emulsion of myth and innovation as we negotiate risks, seek comfort and satisfy our tribal minds.

HANS DOWN

"That is Count Dohna," the horse trainer Wilhelm von Osten said to his companion, Hans, in introduction. Half an hour later, they stood before a blackboard that had been laid flat on the ground, an alphabet scribbled across it in white chalk. Wilhelm pointed to the count and asked Hans if he could remember the officer's name.

Hans paused, and then tapped at a letter with his foot. "D" . . . and then an "o" . . .

Wanting to encourage him, Wilhelm started to sound the name out. "Duh . . . oh . . ."

The rest of the letters gradually came as a foot tapped out "h . . . n . . . a." Amazed, the officers and the rest of the crowd cheered. They had just watched Hans correctly factorize four into eight and calculate the sum of five and nine, and now this. It was unbelievable, given Hans was a horse. A horse not only skilled in counting and doing arithmetic, but spelling!

Clever Hans, or *der Kluge Hans* as he was popularly known in early 20th-century Germany, seemed to have no boundaries to his intellectual talents. He could tell the difference between colored pennants, identify objects such as umbrellas and different hats and even distinguish musical tones. Von Osten, his trainer, also happened to be a schoolteacher and a phrenologist. He claimed to have spent years teaching Hans as one would a small child, patiently coaching him to recognize numbers and letters, and showing him how to combine them to answer questions with a tap of his foot. While he wasn't perfect, Hans allegedly had a success rate of close to 90 percent and was said to possess the mathematical aptitude of a young adolescent.

By all accounts, Hans seemed to be one smart horse. Thanks to Darwin's popularizing of the concept of evolution, the public was increasingly intrigued by the notion of a distant but real family relationship between humans and other animals. Intelligent horses were just one more curiosity that captured their imagination. Some, including Germany's Board of Education, however, weren't quite so convinced that Hans was the real deal.

In 1904, the psychologist Carl Stumpf formed a commission of investigators who were charged with determining the authenticity of Hans's abilities. Among the investigators were a veterinarian, a circus manager and several teachers. Stumpf appeared to suspect von Osten of deceit and proposed that Hans was being deliberately cued to tap answers he didn't implicitly understand. Surprisingly, the investigators' final conclusion was quite definitive—they could find no evidence of trickery. It seemed von Osten was not attempting to dupe his audiences, and therefore Clever Hans was, indeed, quite clever.

Following the investigators" report, the evaluation was passed onto another psychologist, Oskar Pfungst, who perused their methods and felt there was still some room for doubt. Conducting dozens more trials, Pfungst varied a number of the factors involved in the horse's demonstrations by putting blinders on the animal, having different people ask the questions and by hiding the questioner from sight whenever he asked a question. Interestingly, the horse would tap out the right answer regardless of who was asking it.

But Hans wasn't quite the equine genius people had imagined. If the questioner knew the answer, Hans would respond correctly nine times out of ten. If the questioner didn't know, or had the wrong answer, Hans's success dropped to a few percent. Either Hans was a gifted mind reader, or there was something else going on. When the questioner was hidden from the horse's view, his talent once again vanished. Hans had to be able to clearly see the person who was asking the question.

Pfungst agreed that von Osten wasn't intentionally deluding his audiences. That wasn't to say Clever Hans was a genuine prodigy, however. Rather, the psychologist concluded that questioners were subconsciously using their bodies to cue Hans as he went through the process of counting out an answer. Once the horse reached a spot that was presumed to be correct by the questioner, the questioner's posture would change just enough for Hans to notice. This would become known as the Clever Hans effect and is regarded today as a significantly influential factor that needs to be taken into account any time a test involves an interaction between people or animals and an experimenter.

Most horse wranglers know how horses communicate with one another through rather subtle changes in their stance and posture. For a horse, Hans was no dummy—he'd learned to read a human's body language for a reward of sugar cubes or carrots. While inarguably impressive, his brain wasn't quite able to contemplate abstract concepts involving fractions or sums. Nor could it translate sounds into phonetic symbols such as letters. Hans appeared clever because he had learned to associate reward with physical cues, not because he truly comprehended what he was hearing.

We're all clever horses, in our own way. Our neurological machinery makes it easier to act in response to the cues and opinions of others rather than tediously evaluate every piece of information we receive as an individual learner.

Termed "social epistemology," we humans have a tendency to create knowledge collectively rather than as individuals. While this can occur in many different ways, social learning theory indicates we learn by observing others and recognizing when they are rewarded or punished as a consequence of particular behavior. By the same token, voicing a belief that is regarded by most as unpopular could earn a dose of mocking or even exclusion from a community, which makes it more difficult to critically evaluate a belief than to simply go with the flow. If an idea runs against the values of community members, it is far more easily dismissed than if it appeals to their sense of morality.

The very choice of words used to describe an idea influences not just our perception of that idea, but how we see the world. The early 20th-century American linguist Benjamin Lee Whorf suggested that the language we think in might constrain our thinking habits. An interesting example is the Kuuk Thaayorre people of north Western Australia, who have no words for left and right. Instead, relative positioning of an individual is from the perspective of the compass directions. In English, if you wear a glove on your right hand and then turn to face the other way, it's still on your right hand..Speaking in the native tongue of the Kuuk Thaayorre, however, the glove would be on your western hand before turning around, and then your eastern hand after doing an about-face. If you have trouble remembering your left hand from your right, imagine if you had to keep track of a mental compass! Interestingly, this is precisely the effect such a thinking habit seems to have. Kuuk Thaayorre linguists appear to have an above-average appreciation of absolute spatial dimensions.

On a more subtle level, the words we choose to use in describing an idea can make the idea more or less palatable. Nobody knows this better than the organization People for the Ethical Treatment of Animals (PETA), whose website proudly describes fish as sea kittens, because "who could possibly want to put a hook through a sea kitten?"

In telling stories as a community, we establish and reinforce our beliefs. The nuances of the language we use often convey more than absolute facts as pure qualities, carrying cultural baggage that can actively shape how we think. It's difficult to consider how a fish might feel pain without empathizing from a human point of view, turning a neurological response into an emotion-charged event. Try as we might, our storytelling heritage intrudes and contributes a personal perspective.

Our perception depends largely on a diverse range of beliefs and thinking tools that we inherit from those we consider to be respected members of our tribe, gained subtly as we observe the rewards and punishments of others as they share ideas. In fact, when it comes to picking and choosing our beliefs, it's often all in the memes.

MEMING OF LIFE

A rather perplexing problem troubled embryologists during the 19th and early 20th centuries: if life was just a complicated web of chemical mechanics, how could a single fertilized cell divide into two, two into four, four into eight . . . and so on to produce an adult organism with a variety of different tissues and organs? If cells were just chemical machines, what sort of machine could keep breaking in two and yet continue to function?

A suggestion came from a man better known for his work in physics than in biology. In 1944, the physicist Erwin Schrödinger wrote the book *What Is Life?* based on a series of lectures he had delivered at Dublin's Trinity College. He argued that it was possible for chemistry to be solely responsible for coding all of the information needed to create a fully functioning organism, and that no mystical, undiscovered field of science needed to be invoked. The code would necessarily involve some sort of solid material, or "crystal," made up of a complex pattern of non-repeating units.

Within a decade, molecular biologists James Watson and Francis Crick had published a model of Schrödinger's so-called aperiodic crystal, basing their work on an X-ray diffraction image taken by the physicist Rosalind Franklin. DNA, or deoxyribonucleic acid, proved to be Schrödinger's chemical material that transferred information from one generation to the next.

Genetics has become a useful theory explaining why you resemble your parents and not your neighbors, your cat or a blue whale. As we learn more about the chemistry responsible for our inheritance, our understanding of evolution, diseases and medicine improves.

The anatomy and biochemistry of the human species don't seem to be the only things that have changed with time. Our culture has as well—language, technology, entertainment, traditions, superstitions . . . all manner of human behaviors appear to evolve over the years and generations. Of course, some of this could itself be the result of genetic differences, evolving as our brain changes over numerous millennia. But a great deal of behavior changes far too quickly for genetics to be the sole cause. Evolution can't explain why I won't be wearing a stove pipe hat and a vest to the cinema any time soon, or why tattoos were once for scruffy-bearded outlaws and can now be found on the arms of upper-class lawyers, teachers and even politicians.

Some biologists and sociologists recognize that there are similarities between changes in culture and the evolution of an organism. German biologist Richard Semon's 1921 book on social evolution, *The Mneme*, was named after the Greek goddess of memory. Semon believed that society evolved much like a population, with "mnemes" transferring observed behaviors into physical features of the brain. Yet it was the Oxford professor and popular science writer Richard Dawkins who became responsible for the word's popularity today, independently arriving at a similar spelling by combining the words "gene" with "mind," or "mimicry."

Defining a meme in broad terms is quite simple—it is any idea or behavior that can be communicated from one mind to another. Just like genes, memes can be miscopied or "mutated" as they are transferred. No form of communication is perfect, leading to variations in behaviors or ideas. Also like genes, some ideas survive while others perish, given some behaviors work better than others in different environments and therefore lend themselves to being copied by other people.

Words are prime examples of memes. Adolescents love to share words that have rather exclusive meanings within their peer group. The words' usage identifies who is part of their tribe and who isn't. A catchphrase, or even a meaningless sound, might be inherited innocently from a television program or an amusing experience at a party. Use of this phrase or sound provokes a reaction from others, providing feedback that changes how the phrase or sound is interpreted. Others might hear it being used and then copy it themselves under similar circumstances. In this environment, the word proves to be a useful way to bond with others within the peer group. Finally a teacher uses it, dramatically changing the social environment. No longer is it exclusive to the tribe of students, used to identify the "cool" clique—the expression has swiftly lost its charm and is abandoned in favor of new words or sounds.

This might make perfect sense for simple behaviors like words or style of clothing, but can memetics describe far more complicated behaviors such as religion?

African slaves transported to the Americas brought with them a rich culture of dance, song and ritual. Openly practicing their heritage, however, was typically forbidden. Consequently, the slaves modified the rituals and dances of their ancestral land in order to preserve something of the community's culture without compromising the rules of their new social environment. Deities from the old world took the form of saints in the new pantheon of gods. Songs and dances were performed in the context of Christian ceremonies. In other words, memes such as song, dance and language inherited from previous generations needed to prove useful (and not detrimental) to a new environment if they were to continue to survive, forcing them to mutate. Today, America celebrates a rich musical culture, comprised of jazz, blues and gospel choir music, that evolved from behaviors imported across the Atlantic in the minds of generations of African slaves and adapted to suit a new social landscape.

What of science, then? Can the philosopher's toolbox be described as a collection of memes? Can an appreciation of thinking tools such as blind testing, logic, empiricism and probability be copied between individuals? Do people inherit scientific values and ideas from their social group as they would a religious value, a cliché, a dance, a catchphrase or a musical taste? Perhaps. Science is certainly a behavior that has evolved over the centuries, with old practices falling aside when they no longer seem to be useful and new ones being shared when they appear to work. Science has evolved as a methodology, and will certainly continue to change as time goes on. Fashions come and go in laboratories just as they do in music and art.

Memetics is not a concept favored by everybody. A common criticism is that, unlike genes, memes have no physical unit of transfer. Scientists can identify a gene by the sequence of chemical bases in a strand of DNA. How do you define the precise limits of a single meme? Is it a specific arrangement of neurons? A particular brain function?

On the other hand, as proponents of memetics like British science writer Susan Blackmore point out, describing the inheritance of physical traits according to a system of genetics wasn't completely useless prior to the discovery of DNA. Not only that, geneticists are slowly starting to abandon the "one gene, one trait" concept of inheritance. Attributing a simple sequence of chemical codes to a single gene is not as handy as it was once thought.

Memetics might help us to understand how the general public

(including scientists themselves) understands and appreciates complex social behaviors such as science. Looking at the philosopher's toolbox as a collection of values that can be copied, changed and passed on as a culture could help us understand not just its limits and its benefits, but how to help others better use the toolbox.

We all inherit vastly different combinations of memes depending on which tribes we associate with. Our family and friends might hold strict religious views, and yet we might have work colleagues or other social groups who value logic and the consensus of a different, non-religious community. We face a constant tug-of-war of viewpoints that makes it difficult to know how to determine good ideas from bad ones.

Since the 19th century, the "scientific" meme has strengthened, spreading on account of its demonstrable pragmatism and ability to accurately predict the future. Yet it continues to compete with values that are better reinforced by our social minds. Not only do we still see the face of Christ in spilled paint, we now feel it must also be validated by scientific facts. If they conflict, the meme that works best in that person's environment will win. Often that means dismissing logic or an internally consistent model of our universe and succumbing to the strong desire to believe in a god who appears to the faithful in mundane ways.

Those critical values and ideas that have developed throughout the ages—the philosopher's toolbox—face an uphill battle spreading through a community. A person who appreciates logic might not always know how to identify it. Statistics might be a good way of sorting information, but our brains aren't calculators. Our brains are also memetic environments, and favor some ideas over others depending on how it benefits us socially.

SELECTIVE MEMORIES

I started my working life many years ago as a medical scientist in a clinical laboratory. The job meant covering night shifts in a hospital, analyzing blood (and any other body fluid you can name) for the accident and emergency and intensive care departments. Some nights were a blur of lab tests and phone calls as overdoses, chest pains and busted appendices streamed through the doors, while others were eight hours of relative silence. Once a month, however, as the full moon rose over the hospital, at least one person—be it a nurse, a doctor or a paramedic—would cautiously predict that the night would be busy.

I never kept a record of those nights. At least one evening, from memory, was particularly chaotic, and it stuck in my mind solely because the doctor on duty kept mentioning that the full moon brought out the "lunatics."

The fact that the very word lunatic comes from the Latin word for moon indicates that this association has been around for some time. Bright nights belong to werewolves and madmen, traditionally speaking. Strangely, flicking through numerous records for various hospital admissions and police arrests fails to reveal any moon-related pattern. Rather than just being an act of mixing up correlation and causation, it's a case of no correlation existing at all. A review conducted in 1986 by the psychologists Ivan Kelly and James Rotton and the astronomer Roger Culver failed to find any connection between the full moon and a rise in not just hospital admissions, but episodes of domestic violence, crisis calls, suicides or even positive events such as casino wins. In other words, that same doctor who swore that it got busier on one particular evening each month probably wouldn't be able to pick the dates of full moons from a list of patient admission numbers.

Sometime in our past, a correlation between abnormal behavior and the brightness of the moon was assumed. Perhaps it had something to do with monthly feasting and revelry; a bright, moonlit night is certainly better for parties. In any case, this association has been passed down through the generations. Old nurses tell new nurses, who then share it with medical scientists, who tell their husbands and wives and so forth. The young doctor might well have heard about it from a colleague or during his studies.

Surely if there was no effect, we'd notice . . . right? If there is no evidence of an increase in hospital admissions, why would the doctor believe otherwise?

Like the rest of us, the doctor's nervous system has a talent for sorting information. Rather than waste time and energy holding onto every scrap of information that enters through his senses, his brain retains only the snippets that it deems to be relevant at that moment.

In this case, his brain paid close attention to patient numbers during a full moon and flagged information that supported what he already presumed to be true. On nights where the numbers failed to support the belief, the doctor would simply forget that it had been a full moon.

Described by cognitive psychologists as "confirmation bias," we have a tendency to passively favor observations that support what we already believe to be true, rather than actively search for reasons to contradict it.

Here's a quick example, a test created by the cognitive psychologist Peter Wason that shows what happens when we do want to test a hypothesis.

Imagine four cards on a table, labeled as per the following:

E K 4 7

Now, consider this: "How many and which cards do you need to turn over to judge whether the following rule is true: *'If there is a vowel on one side, then there is an even number on the other side'*?"

What would you answer?

It turns out that most people would either say one, and just turn over the E, or two, flipping the E along with the 4.

Choosing the E is a good move, as an even number on its flip-side would help support the rule "*if one side = vowel, then the other side = even*," while an odd number would refute it. But turning over the 4 won't tell you anything. The wording of the rule refers specifically to vowels, not consonants. If the 4 has a vowel on the back, the rule remains confirmed . . . but if it has a consonant on the back, the rule is still possibly true.

By flipping over the 7 and finding a vowel, on the other hand, the rule has been broken.

Only 4 percent of people flip over the E and the 7 together. Confirmation bias is the tendency for people to consider ways of confirming the rule rather than actively seeking reasons to break it. The doctor who believed that patient numbers went up during a full moon would probably not have been paying close attention to those evenings where admissions were down, as that information failed to match his expectations and therefore wasn't retained by his memory.

There is no clear explanation for why our brains do this, however, it could simply be yet another consequence of our brain gambling to save energy.

Psychologists use the term "heuristic" to describe a short cut your brain takes in creating a mental instruction list for certain tasks. You might determine which cafés have tasty food based solely on the look of their window display, or assess whether a movie might be worth watching based on the director or cast list. We all draw on aspects of our experiences and use clues such as aesthetics, popularity, costs or a combination of qualities to decide what action we should take. Walking through a dark vacant lot behind a stranger who stinks of alcohol might prompt fear, thanks to a heuristic that acts on a stereotype you've formed of dark alleyways and intoxication. That same heuristic probably won't cause you to quicken your pace when walking in front of a little old lady.

Using a heuristic that identifies information you're already familiar with is far more efficient, and often more useful, than actively searching for signs that might contrast with previous experiences.

Another possibility is that confirmation bias arises out of our motivation to support commonly held beliefs rather than refute them, which is simply another function of brains that evolved to be social machines. Believing in the lunar effect with your family could once have been considered to be far more beneficial than risking isolation by speaking out against it.

Whatever its origin, confirmation bias has an extremely strong influence on how we make sense of the world. Nobody pauses to consider the millions of mundane, boring encounters they have ever had in their life. The events we pay attention to are the freaky phone calls we get from an acquaintance we were just thinking about that morning, conveniently ignoring the countless occasions we've thought of somebody and then not heard from them.

Thankfully, confirmation bias is easily handled by the philosopher's toolbox. Properly blinded tests make it impossible to subconsciously match a test's results with prior beliefs. By systematically listing each day's hospital admissions without marking it with a date, an experimenter can create a virtual blindfold that prevents them from associating some of the numbers with phases of the moon while ignoring others. Unable to confirm their belief by only identifying those full moons with higher numbers of inpatients, the experimenter is forced to view all of the information together in order to come to a conclusion.

Yet as simple as this solution sounds, even the most robust blinded experiments can produce a counter-intuitive or even downright bizarre result. The question is: do you accept the strange data or risk confirmation bias by dismissing it as an "obvious" error?

A DRAWER FULL OF RIPE CHERRIES

Canadian psychologist Kevin Dunbar is a scientist's scientist, studying how researchers go about performing their investigations. In the past, science philosophers have offered a range of opinions on how science should ideally be practiced, whether by giving a nod to

Francis Bacon's scientific method, applying Karl Popper's criterion where science is defined by an idea's ability to be disproven, or their own preferred set of criteria. But Dunbar wasn't so concerned with how science should be done in a perfect world; he wanted to know how modern science is actually practiced in the laboratory in everyday life.

In over half of the situations Dunbar observed in his study of laboratory practice, the results of the scientists' experiments took them completely by surprise. Occasionally their data made no sense whatsoever. Far from being a torch for illuminating the dark corners of the unknown, the act of "doing" science was like groping around for the light switch and constantly finding the kitchen sink. What was thought to be the exception proved to be the rule—the results produced by even the most meticulously designed experiments were rarely as expected.

It's amusing to think that Dunbar might have been surprised by his own findings. In any case, strange data are hardly anything to worry about. In the words of the science fiction author Isaac Asimov, "The most exciting phrase to hear in science, the one that heralds new discoveries, is not 'Eureka!' (I found it!) but 'That's funny. . . .'" What was of concern to Dunbar was how scientists typically dealt with those unexpected results.

Initially the scientists would go over their method with a fine-tooth comb in search of any mistakes they might have made. Apparatus would be swapped, chemicals replaced, procedures checked, fuses repaired and socks and underwear changed. However, in cases where no error could be found and the results were stubbornly persistent, many scientists simply put the whole experiment to one side and ignored it. Rather than treating their strange find as grounds for some amazing new discovery, anomalies were being seen as inconvenient dead ends. This wasn't the science of Hollywood and comic books—it was the science of checkbooks and deadlines.

Or perhaps it's got more to do with our picky neurons. This behavior appears to be the work of a small part of the brain up the front of your skull called the dorsolateral prefrontal cortex (DLPFC)—also known as the brain's editing suite.

To determine precisely what was influencing the scientists' behavior, Dunbar tested a handful of students by showing them two videos of falling weights and watched their brains work by using an fMRI machine. To a physics student, the idea that two different weights will drop at different speeds is ludicrous. On the other hand, students who didn't know any better might be expected to intuitively reason that lighter objects should fall slower than heavier ones. On being shown the video of two different masses falling at a different speed, an alarm went off in the physics students' heads, alerting them to the fact that something wasn't right. This didn't occur with the non-physics students, who were obviously quite relaxed about the laws of physics being broken right before their eyes.

Dunbar wasn't quite done yet, as the brains of the physics students had yet another trick up their sleeve. On seeing the impossible footage, their DLPFCs went to work labeling the new information as irrelevant and, therefore, to be ignored. In effect, the alarm bells and the brain's editing suite prevented the peculiar information from sticking in their memories.

The DLPFC takes time to develop in humans, maturing only in late adolescence. Prior to this, children and young adults have trouble paying attention thanks to their inability to focus on relevant stimuli. However, when this section of the brain eventually functions, it's responsible for discriminating between information that is highlighted as relevant and information that is seen as a waste of time retaining.

Often, the act of finding relevant information to support what we already believe is much less subtle. "Cherry picking" is a term used to describe confirmation bias when it comes to finding particular reasons to explain our beliefs. Like a fruit picker selecting the most delicious looking morsels from the orchard, it's easier to point out research that puts your belief in a favorable light while ignoring or creating excuses for experiments that don't. This might not always be a conscious choice, but rather a function of our brains weighing all of the information we come across and determining when something is important and therefore memorable, and when it is unimportant and therefore potential mind-clutter.

A related behavior to cherry picking is given the rather ominous title of the "filing drawer." If an organization wishes to show people their product is the best, or even simply safe, they can conduct an experiment. A university department famed for a particular discovery might also conduct a follow-up experiment. Should those studies produce favorable results, the results will readily adorn the packaging of the particular product in bright, bold colors and probably become a slogan. But what of those experiments that don't go quite to plan? What if the experiment shows the product is faulty, or that the original discovery was wrong?

The filing drawer is like a dark orphanage for experiments that don't turn out as hoped. They are judged more harshly, meaning more of them never see the light of day, hidden from discussion because the results weren't to the experimenter's liking. Because nobody gets to see them, it's impossible to know a great deal about their results.

Given the benefits of learning from the mistakes made by others, the filing drawer and cherry picking represent important biases in how science is done. If science were indeed a single organization of like-minded people, all agreeing to keep secrets and getting on perfectly well, cherry picking and the filing drawer would be far more sinister issues.

Fortunately, the scientific world is far from a single, all-embracing tribe. Its members come from different backgrounds and often have different influences. As a result of these differences scientists fight, compete, argue, express a range of personal biases and generally want to be heard and recognized for their work. The only thing that remains detached and consistent throughout these academic gladiatorial matches is the universe the scientists are focused on. Eventually a researcher who is not affected by the same social forces as previous generations repeats the experiment, only this time without the restrictions that once prevented it from being brought out into the light.

To borrow a phrase from Canadian science fiction author and biologist Dr Peter Watts: "Science doesn't work despite scientists being asses. Science works, to at least some extent, because scientists are asses."

Science isn't pretty and it's not simple. It is messy. Most importantly, it takes time to gradually move on. But this slow progression of ideas is important. Scientific hypotheses suffer a Darwinian evolution—survival of the fittest—where useful ideas will eventually persist and ideas that fail to live up to the same standards, ideally, should fade into history.

LITTLE MISS MEMORY

What memories do you have of August 31, 1997? The date itself might not be as remarkable as September 11, 2001; however, once you associate it with the death of Princess Diana it's possible you'll suddenly recall with crystal clarity precisely what you were doing. Likewise, on mentioning the passing of Michael Jackson or a momentous occasion such as the 1969 Apollo moon landing, there's a good chance (if you were alive then) that it would spontaneously summon a flood of images.

I was sitting in the staffroom of a fast food restaurant, awaiting the start of my early morning shift as a kitchen hand. To this day I can remember holding my apron in my lap, flicking through a newspaper. A manager walked in and asked me if I heard that Princess Diana had died in a car accident. For some reason, my initial reaction was to dismiss it as a joke. The manager's face—serious and earnest—is clear in my memory as she strove to convince me otherwise.

As others have recounted their "flashbulb memory" of that sad occasion, that's the memory that has come to mind for me. Unfortunately it's incorrect. Completely. It has to be. I could not have had that experience; when I was collecting information for my curriculum vitae several years later I realized I didn't actually start work at that establishment until the middle of September, a good week or two after the event. How could I have such a vivid recollection of something that never occurred?

We all consider memory to be a faithful mental reliving of an event, as if our brain stores our sensations in a neurological hard drive, and we simply press "play" to see them again. As comforting as this thought is, there is a wealth of evidence suggesting our memories are far more malleable than we'd like to think.

The American psychologist Elizabeth Loftus is known primarily for her numerous studies into the reliability of memory. In the 1970s she investigated the effect subtle suggestion has on our recollection of a recent event. Volunteers in one particular study were shown a slide of a car pulling up to a street sign. For some, the sign clearly said "stop." Others saw "give-way" (or "yield" in the United States). On questioning, however, experimenters referred to the sign the volunteers did not see, swapping the stop sign for the give-way sign, and vice versa. A significant number of the volunteers not only failed to notice this twist, they subsequently expressed seeing the false sign. In another part of the study the volunteers were presented with the scene of a car accident, described as either a "crash" or "smash." Those who heard the word "smash" tended to report, in a later discussion, seeing glass fragments that weren't there.

The observations made by the volunteers in this study could not have initially included glass, or the wrong sign. Therefore, their memories could not have been faithful reproductions of their experiences. Even without the influence of a third party, our memories are less than perfect. No matter how confident we feel about the perfect clarity of a past event, it's as likely to be corrupted as one we vaguely remember in hazy details. The more time that passes us by, the more likely a memory is twisted and bent out of shape.

Obviously this would come as a shock to any justice system relying on eyewitness testimony. Yet our desire to mistakenly associate confabulation with lies and inaccurate memories with poor intelligence means we'll readily view a respectable, smart citizen as a trustworthy source of information. In Australia, juries are warned of the dangers of relying solely on the evidence of an eyewitness.

Bad memory initially appears to be something of a heavy impediment for brains such as ours. We rely so much on reliving the past as memories. Yet it all depends on how you look at it. Memories that fail to evolve with time and new experience can be even more of an impediment than ones that we adjust as we learn.

Psychologists Barbara Tversky and Elizabeth Marsh demonstrated our tendency to remodel memories by asking subjects to play the role of a prosecutor in a court trial. Before the role-play they were given a story to read, detailing an unsolved murder and two suspects. The study's participants were then asked to recount a detailed description of one of the accused characters. Some were asked to keep the account neutral, while others were asked to present it as if they were a lawyer summing up for a jury. Finally, they were asked to retell the story they'd read. The results: those who acted like a prosecutor and tried to persuade a jury got more details of the story wrong, mistakenly attributing details of one suspect to the other while remembering more incriminating details.

Each time we re-experience a memory, we make adjustments in light of other more recent memories. Rather than an album of snapshots, our mental past is a collage we cut up and glue back together constantly. While my manager could never have revealed Princess Diana's death to me in the kitchen of a fast food restaurant, aspects of the memory are probably accurate. That manager actually existed and was friendly and humorous. I read the newspaper before my shift most mornings. There might have been talk of Diana's death in the following weeks, or perhaps a memorial service was shown on the television in the back room and the manager might have commented on it. As time passed and I was encouraged to consider that moment in time, my brain probably seized on several clues and reconstructed the entire affair using what information I had gathered since. It was only when I encountered a dissonance between the memory and factual records of my work history that I was forced to consider I might be wrong.

As sad as Diana's passing was, if I'd felt it as a personal tragedy it's possible I might have had a better chance of accurately recalling more details. Negative emotions, according to Boston College psychologist Elizabeth Kensinger, appear to sharpen our ability to remember certain details far more faithfully than positive emotions.

While memories are great things to ponder in our old age, they're more useful as guides that influence our behaviors than as mental movie marathons. A jury might like an eyewitness's memory to represent real events and their confidence to be a reflection of its accuracy, but the truth is we never evolved to have picture-perfect memories, but ones that are updated every time we think about them. Collectively, as we share information, this could be rather beneficial. A single experience by a lone individual risks being an outlying event, unrepresentative and potentially misleading. But memories that shift and change with the experiences shared between numerous individuals might have a better chance of leading to behaviors that have an improved chance of reflecting a generalization of nature rather than a specific event experienced by any one mind.

WHAT ARE THE CHANCES?

Las Vegas is a city built by losers. There is no more beautiful, perfect example of the brain's ineptitude when calculating risk than this colorful kingdom of casinos.

Probability, or chance, is a personal estimate of the likelihood of a particular outcome based on limited information. For instance, I'm not confident that my sister is going to visit me today, given she lives 1,200 kilometers away and has a habit of planning such an event well in advance. What if she did ring to say she was on her way? I might be surprised, but my confidence in that event occurring would shoot up to "near certain." On seeing her step off the plane the event would be 100 percent certain—she'd have arrived, meaning I wouldn't think of it as a probability anymore, but an actuality.

In effect, probability is only a reflection of how much or how little we know of a given situation. Since tomorrow hasn't happened yet, I don't have direct information of future events. Will the sun rise? I hope so, and based on a mix of past experience and theories about the solar system I have great confidence that it will. But while I know the sun appeared over the eastern horizon this morning, I can only be relatively less than certain that it will happen again in another 24 hours.

We calculate most risks on the fly without much consideration. Most of us will make breakfast without wondering if the food is poisoned, put our shoes on without thinking there might be scorpions in them and turn on the television or computer without snatching our hand back for fear of electrocution. All it takes is one sick morning, a bite on our big toe or a shock from a short circuit to make us hesitate and recalculate the odds.

Strangely, each half of your brain performs this task in its own way.

Consider a situation where you're asked to pull colored tokens from a sack. You're told that three-quarters of them are black and the rest are white. Before you pull out a token, you're asked to guess its color. If you think about it, there are two ways of going about this.

You could just say black each time, and know you'll be right three out of four times. This is called "maximizing." Unfortunately, you can also guarantee you'll be wrong 25 percent of the time. The second method is called "frequency matching"—you can mix it up a little, answering black most of the time and throwing a white response in every now and then. While there's a chance of getting them all correct, there is also a risk of guessing correctly less than 75 percent of the time.

Want to know which is more efficient? Again, consider Las Vegas. Casinos don't vary their guesses based on whether they think they'll win or not on each bet. The odds are already stacked in their favor. Sure, some lucky punters will walk away with a few million, but by maximizing the casino knows they'll win far more than they could ever lose. Like the casino, the right hemisphere of your brain has a tendency to maximize the odds.

But your left hemisphere prefers to frequency match. The reason it does this might have something to do with its ability to find useful patterns in the chaos of everyday life. By combining patterns, we're able to identify potential cues that can better inform us on which odds are in our favor and worth investing more in. That is, of course, assuming that there are any useful patterns in the first place.

Casinos will encourage punters to find possible patterns by leaving pencils and small score sheets at various tables. The fact that such patterns don't exist doesn't dissuade people from thinking they've unlocked the game's secrets. For instance, it's not uncommon to find gamblers standing near a roulette table, recording the square the ball lands on during each spin.

The casino knows all too well that there is no pattern that will allow a person to predict with any accuracy where the ball will land next. That doesn't stop determined gamblers from trying, of course. Faced with a run of 15 odd numbers, most people would find it hard to avoid believing the next number doesn't have a higher chance of being even. Without knowing the precise amount of force placed on the ball as it spins into action, the acceleration of the wheel, the air resistance as it bounces from number to number (or a range of other factors), your only option is to say that the ball has less than a 50 percent chance of landing on an even number. It's just as likely to land on an odd number a 16th time for all we know . . . or land on that dreaded "double zero."

In fact, the only variable that has absolutely no influence on where the ball might land is the number it landed on previously. So what convinces people that there is a pattern they can predict?

Imagine a string of 50 numbers—odd, even, even, even, odd, even . . . and so on. There are runs of odds and runs of evens. Now, picture 100 different strings of these numbers.

Most of those strings will consist of a jumble of relatively short sequences of odds and evens. Very few (if any) would be made up of 50 odd numbers or 50 evens. This is what a gambler focuses on when they search for a pattern—they are gambling on the belief that those runs of odd or even numbers are a pattern. The brain is a pattern-matching machine, and most patterns do indicate something useful. A roulette wheel that keeps coming up with long strings of even numbers or doesn't ever come up red might be a bit suspicious.

Unfortunately, the gambler only has so much money. While long runs of evens or odds or reds or blacks are rare, what matters is the chance of guessing correctly for each particular run versus the amount of money the gambler is betting. On each and every spin, the gambler is trying to match the frequency according to a relationship they mistakenly believe influences where the ball will fall next. Meanwhile, the casino has maximized—the longer the gambler plays, the more likely it is they'll find themselves handing over their last dollar, confident the streak of odds must be broken. The only rule a gambler ever really needs to know is "quit while you're ahead."

Of course, this probably all makes sense to you when worded like this. Framed in the right way, gambling seems like a bad way to lose good money. Yet context is vital for decision-making. The same information framed in different ways alters how we make our choices.

In 1979, cognitive psychologists Daniel Kahneman and Amos Tversky set out to test a problem in psychology called the Allais paradox.

Think about the following two options:

Option A: You have a 100 percent chance of receiving $1 million.

Option B: You have a 10 percent chance of receiving $5 million, an 89 percent chance of receiving $1 million, and a 1 percent chance of receiving nothing.

Which of these two options would you take?

While B offers more money, most people choose A—a certain win. We have a tendency to focus on avoiding the tiny risk of getting nothing rather than comparing which option contains a bigger reward, at least according to this example.

Now, consider another two options:

Option C: You have an 11 percent chance of receiving $1 million and an 89 percent chance of receiving nothing.

Option D: You have a 10 percent chance of receiving $5 million and a 90 percent chance of receiving nothing.

Most people will choose D over C, going for $5 million even if there is a greater risk of getting nothing. In this case, rather than being influenced by the chance of winning, the decision seems to be based on the size of the reward. Unlike the first two options, the small difference in the risk of walking away empty-handed is now ignored in favor of the size of the win.

Kahneman and Tversky felt the psychological theories of the time couldn't adequately explain people's choices when it came to predicting how we deal with risk and reward, and set out to develop a theory that could take into account the differences in how we deal with risk when it comes to potential gain and potential loss.

Their solution, called "prospect theory," describes our decision-making in two stages. In the first stage, we simplify the situation by roughly ranking the possible outcomes according to what we stand to win and what we stand to lose. In other words, we frame the situation in terms of getting something or losing something.

We then look at our chances, but with a slight bias based on whether we see it as a loss or a gain. Given an option of being given $1,000 or flipping a coin on the chance of getting $2,500, we typically choose the $1,000 and avoid taking a chance. However, if we were facing a fine of $1,000 or a coin flip on a fine of $2,500, we'd view the chance differently and go with the coin flip.

The perception of certainty can make a big difference in whether we choose to take a risk. In the second set of options, the fact that the odds are so similar turns the attention to the $5 million prize. We therefore dismiss the risk factor and focus on what we might win.

Prospect theory is occasionally referred to as loss aversion theory. It is the bias in our minds toward avoiding losing what we already have.

A concept related to prospect theory, as shown in the Allais paradox, is called the "pseudocertainty effect." It predicts that people will avoid risk where the outcome is framed as positive and we seek risk where the outcome is framed as negative.

Imagine a scenario where 600 infected people will die if they're not given a cure.

Cure A will save 200 of them.

Cure B has a one in three chance of saving everybody, but a two in three chance of saving nobody.

Which would you choose? Nearly three-quarters of people prefer Cure A.

Now, think about it this way:

Cure A will see 400 people die.

Cure B carries a one in three probability that nobody will die and a two in three chance that nobody will survive.

Just over three-quarters of people surveyed using this framing of the very same scenario chose Cure B instead. The way we deal with risk is heavily dependent on whether we think we stand to gain or lose more than we currently have.

Another influence on risk is our perception of really, really big (or really, really small) numbers. Having one mouse in your kitchen versus a thousand is pretty dramatic. One is a pest, the other a plague. Yet compare a million mice with a billion mice crawling through your house, and the idea becomes a little overwhelming. Both amounts are considerably huge, even though a billion is a thousand millions.

Whether it's atoms in a pinhead or stars in the heavens, it's possible for numbers to reach a size where they're beyond comprehension. Try as we might, we cannot honestly fathom the distance to the edge of the universe in kilometers, let alone the time it would take to drive there in a car.

(For the curious, a rough estimate of the distance between Earth and the edge of the universe is 91,700,000,000,000,000,000,000 kilometers. It would take 105,751,745,000,000,000 years to drive there if you stuck to a 100 kilometer-per-hour speed limit. For comparison, the universe isn't much older than 14,000,000,000 years by conservative estimates. If you started driving when time began, you still wouldn't make it if you had another million universe lifetimes in which to drive.)

Lotteries rely on the confusion of really big numbers to get you to play. Winning a few million sounds like a dream come true, right? As they say, somebody has to win it, so it might as well be you. Some people do this every single week, hoping they can retire after their lucky break. Mind you, few of these people would think twice about getting into a car and driving to work, or even standing outside during a thunderstorm. Yet the average person would stand a much greater chance of losing their life to a car accident or lightning strike than of winning the big one in the lottery.

Between 2004 and 2005, 62 people out of a population of about 20 million died in Australia as a result of exposure to the elements, whether it was succumbing to an avalanche, hot weather, lightning strike or a flood. So why would somebody take active measures to win Powerball in Australia, at odds of about 1 in 27.5 million, and not take measures to avoid thunderstorms? While theories such as prospect theory might help, there really isn't a simple answer. A wealthy adult throwing a few dollars toward a weekly lottery draw isn't risking much, while not venturing out during a thunderstorm might cost them their job. In other words, it's not a simple case of comparing the probabilities.

However, the big numbers also act as a fog, obscuring the true chances of reward. If I gave you a name and asked you to travel to any city in Australia to pick a random stranger on the street in the vain hope of coincidentally matching them with that name, you might think it impossible. For all purposes, you'd be right.

At about one in 21 million, you'd have a better chance of succeeding than winning the big one in Powerball.

At its heart, science is a way of mediating risks. We devote resources to combat the chances of loss and improve our gains. The values that underpin the methodology and use of the philosopher's toolbox each endeavor to give us a way of comparing the likely outcomes, given we are otherwise so poor at it. Ironically, as science improves our lives, a greater percentage of decisions we make contribute directly to our deaths. Fewer people are dying from communicable diseases in circumstances beyond their control, for instance, but according to Ralph Keeney, an American researcher in decision-making, 44.5 percent of premature deaths in the United States result directly from a personal decision. Smoking, avoiding exercise, drug taking, unsafe sex . . . such behavior represents risks that can be reasonably avoided, according to Professor Keeney. By comparison, only about 10 percent of premature deaths at the turn of the 20th century could be attributed to a personal choice.

Paradoxically, it's possible that by living in a world where we collectively make an effort to mitigate misfortune, we personally continue to be confounded by risk.

A HEAD FOR NUMBERS

"Look, it's quite simple," I'm told. My mathematically gifted friend draws a big plus sign on the page and crosses it with smaller bars, creating a graph.

"Here's a grid—the X axis is one dimension, right? You need only one number to describe a dot on that line. Adding a Y axis gives you two dimensions. Two coordinates will give you a point in two-dimensional space."

All well and good. He then points out that the axes were perpendicular to one another. No problem. "So, what about three dimensions?"

he asks, apparently rhetorically as he speeds into an explanation of another axis that I'm encouraged to imagine sticks out of the page. He calls it "Z." I nod, having no problem with the concept of 3D movies. He proceeds to describe a fourth dimension, using time as an example, and my eyes narrow into a squint. I think I get it. Unable to physically wedge another axis onto the graph, he simply dismisses it by adding another set of coordinates to a growing sequence. "(1), (1,1), (1,1,1), (1,1,1,1) . . ."

"Hang on," I say. "What does a four spatial dimensional object even look like?"

Words like "hypercube" are thrown at me, to no avail. Then he proceeds to tell me a story about some ant crawling along a telephone wire that represents folded dimensions, or something or other, and my mind locks up in want of a Control-Alt-Delete function.

Apparently we could be living in a universe made up of many tiny, hidden dimensions "folded" up and secluded away. How do we know this? Because mathematics says it's possible.

Like me, you probably have a bit of trouble picturing anything beyond three spatial dimensions. There are cool pictures you can see that describe how it might work, but given the limitations of our perception, they just look like morphing blocks pretending to have bigger angles than what we actually see in front of us. It's a cheap trick and not an effective one for people with my form of tunnel imagination. I simply can't imagine five, six or eleven dimensions.

Mathematically, it's relatively simple. We merely add another digit onto a sequence. No need to think of a cube that does impossible gymnastics in time and space. It's a number that works with another number to produce a logical outcome.

If big numbers are confusing, the wizardry accomplished with complex mathematics is downright alienating for the average person. Not only are the mind-numbing algorithms, theorems, proofs and equations often written in strange hieroglyphics, the results can often be counterintuitive. Empty space can bend and twist, light can only ever go at one speed (and nothing can go faster), particles can zip along all possible paths before collapsing along the one that suits them best . . . all mathematical discoveries that defy our expectations. Usually, these are evident only in an abstract world of numbers.

In some rare cases, real-world demonstrations are possible. Not that these demonstrations make them any less bizarre.

In order to test special relativity's mathematical conclusions on something called "time dilation," in 2010 physicists from the National Institute of Standards and Technology used atom clocks to examine the effect gravity has on the measure of time. A pair of chronometers was synchronized according to the vibrations of an aluminum atom, which hummed at a frequency of a million billion wiggles per second, giving the clocks an accuracy capable of keeping time within a second per 3.7 billion years. One clock was then lifted by a mere third of a meter, allowing it to experience a slight difference in gravity. Strange as it is to consider, although each clock still measured time in exactly the same way, the seconds measured by one were slightly shorter than the seconds being measured by its partner, thanks to the tiny difference in gravity. The variation was minuscule, equating about 90-billionths of a second difference over a lifetime.

Of course, this makes no sense. Time doesn't depend on where you stand, does it? Apparently it does, and, not only that, the seconds as experienced by your hair are not the same size as those experienced by your toes. The mathematics behind Einstein's work on special relativity was responsible for figuring this out.

Mathematics is one of the sharpest instruments in the philosopher's toolbox. Quantifying the universe and comparing the resulting numbers provides us with insight that could not be achieved any other way. However, its abstract nature and huge numbers often puts the process and its conclusions beyond the appreciation of those who don't share the passion or the knowledge of researchers.

For the majority of the population, these complexities of science are isolating. Unable to interpret the models for themselves or cut through the baffling stream of jargon, confused by the extraordinarily huge (or tiny) numbers and bewildered by the non-intuitive conclusions scientists often reach, most people fall back on the skills that come naturally to their brains to determine what makes good evidence for a scientific idea.

THE NATURE OF EVIDENCE

Simon Singh was once better known for his books *Fermat's Last Theorem*, *Big Bang* and *Trick or Treatment: Alternative medicine on trial*. Since an article he wrote on chiropractic medicine was published in Britain's *Guardian* newspaper in 2008, Singh has come to represent the risks journalists and writers face when dealing with questionable science.

> The British Chiropractic Association claims that their members can help treat children with colic, sleeping and feeding problems, frequent ear infections, asthma and prolonged crying, even though there is not a jot of evidence. This organization is the respectable face of the chiropractic profession and yet it happily promotes bogus treatments.

It was this paragraph that prompted the British Chiropractic Association (BCA) to sue Singh for libel. The presiding judge agreed that the phrase "happily promotes bogus treatments" implied that Singh had claimed the BCA promoted remedies in spite of knowing they were ineffective. In April 2010, Singh was granted an appeal on the court's decision, allowing him to use the defense that his claim was a fair comment and a matter of opinion. That same month the BCA withdrew their action.

Singh was put in an unenviable position. If it were merely a matter of debating why chiropractors believe in the efficacy of their treatments the discussion would be rather straightforward. According to the courts, however, the offending phrase implied that the BCA held no more confidence in the usefulness of their treatments than Singh. If he had been forced to defend a position he claims to have never held, Singh's hands could well have been truly tied.

Despite the BCA's eventual retraction of their claim after Singh won an appeal against the initial judgment, the damage had been done. Science writers (including Internet bloggers and freelance journalists) are now forced to deal with the potential threat of a libel court action for criticizing the work of another, and they must carefully consider the impact their choice of language has when questioning the merits of a scientific claim. This could mean simply refraining from any form of criticism in the first place, given how two different groups of people can associate conflicting meanings with words we might otherwise take for granted as consistent.

At the heart of the problem in this case is the question of what constitutes "evidence." After all, according to Singh chiropractors have not a jot of it, while they themselves contend they have ample amounts.

At the appeal hearing, the court asked of the Queen's Council for the BCA, "What would your case be if, instead of 'not a jot of evidence.' the article had said 'no *reliable* evidence'?"

I doubt that we would be here," was the response.

The entire case, it seems, came down to the question of what defines the very nature of evidence and whether reliability forms a key part of it.

Science is often described as an evidence-based system of inquiry, operating much like a judiciary for the laws of the universe. In plain and simple terms, "evidence" describes an opinion on what form of relationship exists between two or more observations. This might refer to things we see, smell or hear personally, or even a secondhand account of somebody else's experience. If this observation increases or decreases our confidence in a belief, we can technically call it evidence.

This makes science somewhat personal; what increases or decreases my confidence in an idea might not influence yours. Scientists are therefore always arguing about whether or not an observation can be claimed as evidence for an idea. Collectively, they can come to an agreement, but it remains personal even if eventually a consensus forms.

There is no single qualifier that can definitively transform an observation into evidence for all individuals, which is why open discussion is so vital to science. Simon Singh can truthfully claim to not have found one jot of evidence supporting chiropractic treatment. On the other hand, there are numerous observations that appear to have convinced the BCA that chiropractic treatments work well enough for them to be endorsed.

However, if evidence is a subjective term like "beautiful" or "tasty," surely both of them can be equally correct? After all, evidence would then be in the eye of the beholder.

Not quite. While everybody has an equal right to an opinion of evidence, not all opinions are equally likely to be an accurate representation of reality. You might have the opinion that Elvis is still alive based on the evidence of a blurry photograph, and I might have the opinion that he has been dead and buried for some time based on historical news reports I've read. Only one of us can ultimately be correct.

Determining what makes evidence useful depends on how we go about assessing the relationships between everything we observe. For instance, a photograph of a cat could be evidence that I have a pet. On the other hand, if I had a picture of a lion and said it was my pet, you might feel it's prudent to ask a few questions, even though it's also a photograph. Not many people own lions, while it's possible a person might have something to gain by lying about it.

Yet if I showed you a photograph of my pet alien, there'd be little chance you'd simply take my word on it, and you may start considering whether I've had much experience with Photoshop or latex modeling. In other words, the observation of a photograph would be easily taken as evidence for my owning a cat but not as evidence for an alien in my kitchen.

Carl Sagan famously said in his television series Cosmos that "extraordinary claims require extraordinary evidence." In this case, most people have encountered cats as pets before, providing them with a solid amount of empirical evidence. It's unlikely that you would disbelieve my photograph as being a genuine picture of my pet, because you would also regard your observation of household cats as evidence. The same cannot be said for aliens, however, making it a rather extraordinary claim in need of more observations than a mere photograph or a piece of footage.

How we evaluate evidence depends on epistemology—the process we each employ to construct beliefs. This process typically evolves as we develop, modified by our encounters with different tribes expressing different beliefs, which challenge our perceptions. The educational psychologist Deanna Kuhn categorizes methods of evaluating evidence into four levels, which can be associated with our personal development as we learn.

The first level is that of the realist, who views knowledge with certainty and feels that their perceptions represent an accurate reflection of the universe. Most young children are realists, finding it hard to distinguish reality, as it exists outside of their mind, from their mental construction of it.

Sooner or later, we face challenging experiences that encourage us to progress into becoming absolutists, in which knowledge is viewed as right or wrong. As absolutists we see that it is possible for alternative positions to exist in the minds of others, but these positions are always objectively correct or incorrect.

For a minority of people, their epistemological progress stagnates here, describing a worldview that qualifies knowledge in simplistic terms of black-and-white certainty. Most become what Kuhn calls "multiplists," and start to view knowledge as a construct of thinking, usually contingent and therefore often open to negotiation.

Finally, a percentage of individuals will begin to demonstrate an evaluativist epistemology, judging opinions based on the relative strength of evidence according to a set of values and using thinking tools to determine how confident they should be in the validity of a belief.

Progressing through these levels depends largely on the social interactions of the person, whether it's at school or in the family group. According to Kuhn, progress doesn't come as a result of persuasion or lecturing in a belief. You cannot explain to somebody a process of thinking and expect them to embrace an evaluativist epistemology. Rather, it comes with exploring ideas and arriving at conclusions that later appear to be self-evident. Not only do we learn what to believe from our tribe, we learn how to go about forming our beliefs with a bias about what constitutes good evidence. Changing minds, therefore, is far more challenging than simply substituting conclusions or debunking beliefs. It requires a person to change their fundamental epistemology.

Science is associated with a set of values that guides us in determining how to evaluate our observations. It's like a yard-stick for identifying factors that distinguish evidence supporting a belief as solid or tenuous. The philosopher's toolbox has evolved over the centuries to effectively determine which observations are likely to represent something useful and which are the result of our misleading social brains.

THE WOLF IN SHEEP'S LAB COAT

Heather Mills is known for many things—her charity campaigns, for example, as well as her modeling career and being married to (not to mention divorced from) Beatles star Paul McCartney. One thing she is not famous for, however, is her work as a biochemist or gastroenterologist. So when she claimed that meat "sits in your colon for forty years and putrefies, and eventually gives you the illness you die of" in the Observer newspaper in July 2009, it's difficult to imagine she was referring to any peer-reviewed studies she had personally conducted or to her personal experience in the pathology industry.

Of course, that doesn't make her wrong. It does, however, raise the question of how she came to that conclusion, especially given that it contradicts the observations of those who have researched the topic in considerable depth. According to gastroenterologists, enzymes in the small intestine digest meat into its chemical components, which are either absorbed straight into the bloodstream or passed down through the colon. The whole process takes a matter of days, not years, and risk of putrefaction isn't a concern.

Who is correct? Whose opinion is most likely to accurately predict the average person's digestive process—the person who is interested in the process from an academic point of view, or the one who opposes the eating of meat?

Pseudoscience is typically defined as any idea that appears to be scientific, but for whatever reason, isn't. It's also something other people do. Nobody writes "pseudoscientist" next to "occupation" on their CV. Because of the demarcation problem, it's impossible to stake out clear boundaries on what science is and isn't. Consequently, it's hard to determine fake science from the real thing according to a simple, all-encompassing list of criteria. People who think of themselves as scientists think they are doing science, and think they are doing it relatively well. Others might not agree with their opinion, however.

Putting it simply, a belief could be declared as pseudoscientific when it is presented within a scientific context and yet a person's confidence in it is inflated by non-scientific values. The philosopher's toolbox endeavors to limit the influence of false logic and emotional bias, valuing efforts to dissolve the illusions created by our brain's shortcuts. Pseudoscience results from declaring a conclusion to be more valid than is warranted by its method of formation.

This is still a somewhat limited definition. Individually, we can all embrace theories that are scientifically sound for very social reasons. Even robust science can be seeded within an environment heavy in tribal biases. Is it therefore pseudoscience if we believe in the laws of thermodynamics because we like our science teacher?

Our social brains make it remarkably easy for factors such as confirmation bias and the perception of popularity to persuade us of ideas. Coupled with a tendency to associate science (complete with its jargon, statistics and fuzzy-haired men in white lab coats) with trust and reliability, it's easy to be unwittingly convinced by ideas presented in a manner that sounds scientific, even when there are few logical reasons to have much confidence in them.

If the evidence itself is wrapped up in a parcel of mathematical symbols or encoded in jargon, a friendly translator is often needed to explain the relevance. Yet who's to say they know what they're talking about? In a world where communities stretch around the globe and include faces and names we've never personally met, how do we use our biased social brains to pick the sages from the fools? It's for these reasons that ideas can persist through the generations, maintained by faithful believers in spite of conflicting observations or broken logic.

Chapter 6

THE SCIENCE GRAVEYARD

Why do we hold onto bad ideas?

O ur past is littered with the discarded fruit of failed ideas that withered on the vine in want of good evidence. What often begins as an elegant hypothesis capable of solving the most difficult problem can spontaneously disintegrate under a well-constructed experiment or new discovery. Yet within some tribes, beliefs can persist even as others abandon them, kept alive by a sense of hope, fear, disgust or pride. It's the voodoo of our heart that can keep the dead twitching long after they've been buried.

IN AN UNREASONABLE LIGHT

In the summer of 1904 in the French town of Nancy, the respected physicist René Prosper Blondlot agreed to meet a skeptical American colleague by the name of Robert W Wood to convince him of a revolutionary discovery: an amazing form of radiation with properties unlike any the world had ever seen. The world stood divided on this alleged discovery. Believers passionately declared they had no doubt in the existence of this radical find, while the fickleness of the evidence caused others to express grave doubts.

As an apprentice dimmed the lights for a demonstration, Wood took advantage of the room's sudden darkness and secretly plucked an aluminum prism from its place amidst the apparatus, slipping it into his pocket before anybody could notice. Oblivious to his guest's nimble theft, Blondlot sent a current flowing through a vacuum tube, creating a beam of radiation that shot through the space where the prism had been, before falling onto a sheet of phosphorus-coated cardboard.

With the demonstration complete, Blondlot and his team of assistants took the card and pointed out to Wood a series of dark smudges where they believed the radiation had struck. It would have been clear evidence of a radical new discovery in the world of physics, if not for the fact that any connection between the vague lines on the cardboard and a new type of radiation existed only in the minds of the French scientists. Without the prism to split the beam into its components, the faint markings could not possibly have been the result of any remarkable new rays.

Wood's debunking had been a long time coming, and yet for many it was difficult to reconcile such a grave error with the genius of Blondlot. Over the course of several decades he had proved himself an exceptional scientist. He had earned the respect of the French academic community, been elected as a member of the *Académie des Sciences*, and had been awarded prizes for his extensive work on the theory of electromagnetism, showing that the speed of electricity is rather close to the speed of light. German science had only recently given birth to an exciting new field of physics, with Wilhelm Conrad Röntgen's research on electrical currents inside various types of vacuum tube attracting Blondlot's interest. Questions remained on whether the strange new radiation Röntgen had produced—the X-ray— was a stream of particles or a wave rippling through a medium.

The year before Wood's visit, Blondlot had devised an experiment to test whether X-rays could be polarized. If he could demonstrate this feature, he would have evidence that X-rays were a wave like any other form of light.

One of Röntgen's tubes was placed in front of a pair of electrodes so that the X-rays would pass through the current that sparked between them. By changing the angle of the spark and looking out for any variation in its brightness Blondlot would have demonstrated that the light could be polarized, and was therefore a type of wave. Which was precisely what happened.

Encouraged, the physicist decided to experiment by placing sugar lumps and plates of quartz in front of the beam . . . just to see what would occur. Surprisingly, the radiation appeared to veer off course. It's easy to imagine Blondlot's pulse racing as he gradually came to the realization that he was onto something big—it had already been found that X-rays couldn't be refracted by quartz, meaning it had to be some other type of radiation.

Rather than jump to any conclusions prematurely, Blondlot refashioned his experiment to exclude other potential causes, such as ambient light from the vacuum tube. No matter what he did, the observation persisted. Slowly, he lost all doubt that he had stumbled upon something new and he announced to the world his amazing discovery—the "N-ray," named after the city of Nancy.

Dismissing the name as a quaint act of patriotism would be understating its significance in the affair. Europe's scientific community—indeed the entire world of science—was awash with jealousy and rivalry. The English and the Germans appeared to dominate the fields of chemistry and physics, leaving the French to seek any opportunity to even the score. N-rays were far more than just an interesting phenomenon to the French; the discovery represented their country's superiority in a world rife with nationalistic competition. Viewed in such a context, the uncritical acceptance of Blondlot's find (in the French town of Nancy!) by many of his fellow countrymen becomes much more understandable.

French scientists had caught N-ray fever. Experiments of all sorts were done in biology, chemistry and physics, leading to numerous fabulous discoveries. N-rays weren't just another type of radiation like X-rays, it seemed. In fact, they were nothing short of miraculous!

They could pass through any substance except water. Paradoxically, they seemed to emanate from a variety of bodily tissues, both living and dead, and seemed to flow particularly brightly from certain regions of the brain when they were considered to be most active. If they were shone into the eye, they were said to create hypersensitive vision. Parapsychologists jumped on the phenomenon as a potential explanation for a range of psychic and supernatural beliefs, from the observation of ghosts to speaking with the dead. Physicists found some objects could absorb and store N-rays and release them under physical stress. Chemists were claiming to see N-rays being produced by a number of reactions. Once they started looking, scientists all over the country, and then the world, were seeing N-rays everywhere.

Unfortunately for Blondlot, an equal number of people saw nothing. He had his critics from the beginning, of course, as most scientists do. The German physicist Heinrich Rubens, known for his research on infrared radiation, pointed out that many of the claims being made about the rays conflicted with predictions made by the Scottish theoretical physicist James Clerk Maxwell, making them rather questionable . . . if not impossible. In fact, the vast majority of the N-rays' features flew in the face of what had been learned about the theory of electromagnetism. Either Blondlot's discovery was truly revolutionary and would contradict the conclusions of decades of mathematics and experimentation, or it was all one big mistake.

Nonetheless, undeterred by such theoretical expectations, Rubens attempted to replicate Blondlot's experiments on his own. Not a glimmer of this mysterious N-ray could be detected. Across the channel, the British scientists Lord Kelvin and Lord Rayleigh also tried to isolate them, to no avail. Without irony, some such scientists suggested it might be possible that some people were physically incapable of seeing N-rays. Naturally, the French appreciated this possibility, given the proportion of detractors who were foreigners. *Vive la France!*

Not all was so sound on home soil, however. In an article published in Revue Scientifique, the French physicist Louis Cailletet made the rather illogical observation that were it not for Blondlot's reputation as a scientist, he might conclude that N-rays were a delusion. Personally, he had been unable see them, even as others around him assured him with all sincerity that they could. Yet Cailletet's confidence in Blondlot's authority on the matter served sufficiently as evidence, even if his own experience didn't.

One by one, a number of distinguished French scientists made it known they had tried to replicate Blondlot's work and came away empty-handed. The passion of many was equally matched by the confusion of others. Why were some people capable of seeing this radiation, plain as day, while others were completely blind to it?

It was Rubens who suggested to his colleague Robert Wood in 1904 that he should investigate the mysterious N-ray. The American physicist had a reputation for possessing a sharp wit and a keen eye for deceit, having exposed a number of self-proclaimed spiritualist mediums in the United States as frauds. Prior to meeting with Blondlot and experiencing his production of the rays first hand, Wood tried to create them in his own laboratory, once again without a modicum of success.

The meeting made no difference in changing the mind of either party. Even Wood's final act of deceit—surreptitiously pocketing a key part of the experiment and proving the results were all in the eyes of the experimenters—failed to persuade the French physicist he was mistaken. If removing the aluminum prism had any impact on Blondlot, it only provoked indignation. The Frenchman felt as if his goodwill had been taken for granted and that confusion had resulted from his clumsy use of the German language in the absence of a French translator (Wood's French was actually passable, however, he decided to keep this from Blondlot and his assistants, preferring to secretly eavesdrop on their conversations). What's more, an assistant of Blondlot's insisted he had known for some time that the aluminum prism Wood had taken was unnecessary past a given moment, claiming with full sincerity that the rays stayed bent even if the prism was removed.

While the rest of the world gradually abandoned the concept of N-rays, Blondlot refused to give up. Until his dying day, he could not accept he had been deceived. Among the personal documents found in his possession following his passing in 1930 were notes maintaining the existence of his miraculous radiation.

N-rays finally joined countless other ideas in the graveyard of science. Unable to be seen under blinded conditions, the radiation was eventually deemed to be a figment of misperception rather than hard evidence of something new.

Ideas are abandoned for many different reasons. As with N-rays, sometimes there is simply nothing to explain. What initially appears to be a valid observation is shown to be little more than an artifact of the senses. In other situations, scientific theories aren't entirely buried, but evolve in complexity as we make further discoveries with more advanced technology. The microscope opened our senses to a new world of tiny organisms, for instance, forcing the old theory of gaseous miasmas to evolve into the germ theory of disease. DNA allowed Darwin's ideas about evolution to take on a new depth.

Notions of planets, a moon and a sun orbiting a static Earth gave way to an understanding that our world and her sisters followed an elliptical path around our nearest star. Few people continue to think that heat is a physical liquid in the 21st century, or that a material called phlogiston creates fire, or that light ripples through a unique fluid that permeates the universe. The notions that mice spawn spontaneously from moldy straw and maggots are generated by rotting meat (instead of parent flies) are no longer useful given what we know about biology. Deprived of convincing evidence, people gradually lost confidence in many of these ideas, leaving them to fade into the pages of history.

Some ideas take time to fade from sight. Blondlot refused to relinquish his grip on N-rays until the day he died. Why? What could explain such stubbornness in accepting he had been wrong?

Francis Bacon would have referred to his "idol of the tribe" to suggest the original hypothesis gained momentum out of national pride. But even when Blondlot's fellow French scientists steadily came to their senses and found the idea to be useless, Blondlot refused to believe he had made such a grievous error.

Changing the way we evaluate evidence can often be difficult. Blondlot was an exceptional scientist who had contributed heavily to the field of physics. If not for his unreasonable faith in N-rays, he would have gone down in history as a gifted member of the *Académie des Sciences*. Surrounded by colleagues and fellow countrymen who encouraged his convictions, who forgave his errors and praised his authority and intelligence, it would be difficult for Blondlot to see how he could have possibly been responsible for such a mass delusion.

Few ideas can persist when faced with a lack of scientific evidence. Yet sometimes there are greater forces at work. Social thinking can create a need to believe that impedes on our ability to evaluate evidence. It is during those moments that our confidence in an idea is inflated beyond reason, and what should be dead and buried remains alive and kicking.

THE UNDEAD DOCTOR

Succumbing to illness in centuries past often meant coping with far more than the disease itself. For those who could afford a visit from a private physician, treatment included blistering poultices, potions to induce vomiting or lacerations that bled you within an inch of death. Commoners had it even worse—public hospitals were overcrowded refuges of squalor where bed-rest in such damp and fetid conditions killed those who might have otherwise survived the unsanitary surgeries or punishing medications.

Excluding the occasional free spirit, the European medical establishment consisted of academic ranks that all but dismissed the possibility that their esteemed teachers could ever be wrong. So it was that medicine was all too often dictated by dogmatic beliefs in the flow of mysterious living or Earthly forces, the balancing of bodily fluids or vague notions of toxic residues rather than rigorous record-keeping or frank and open discussion.

There was, of course, a great difference between the educated physician who produced elixirs and took pulses and the barber whose unwashed hands shaved chins as often as they sliced boils, pulled teeth or set bones. The melding of doctor and surgeon was a gradual evolution that took many years to occur, and, even then, proceeded always with both eyes on history's authority and nothing on whether their craft truly succeeded. The physician was typically highly regarded and well respected, while the surgeon was a last resort. Without antibiotics or even the slightest regard for hygiene, this was probably for the best.

Biology is a complex thing. Health can come and go on a whim, depending on a number of factors. A person's body can fail even as they feel a surge of good health. Death can come suddenly to a person who is lively and jovial, or slowly to a person who every day looks ready for the grave. Knowing how to fix bodily problems is just as confusing. To the ancient Greek physician Hippocrates, the best action was no action at all. The body should be left to its own devices and heal with minimal intervention where possible.

Yet for most people, this is a difficult thing to do. In the face of potential death few can wait patiently and simply hope for a favorable outcome. Our brain's woeful talent for assessing risk makes it difficult to appreciate the odds of success versus failure. Action feels like a much better option than complacency when one's world feels like it is about to disappear forever.

Hence, physicians plied a trade of draining bodies of their precious fluids, stinging their patient's skin with nettles, feeding them gut-wrenching formulas to make them vomit out any "poisons" and injecting various tonics up their rectums, all to make a dying soul believe all was not yet lost. Doctors had little idea how such remedies might have worked, or even if indeed they did work at all. What mattered was that surviving patients advertised their appreciation, while the failures were quickly buried with a quietly murmured prayer. The good thing about medicine in ages past was that confirmation bias was far more effective if the dead were no longer around to speak of their disappointment.

The German physician Samuel Hahnemann knew all too well how his profession amounted to little more than a blind gamble. In a letter to a colleague he wrote, "The thought of becoming in this way a murderer or malefactor toward the life of my fellow human beings was most terrible to me, so terrible and disturbing that I wholly gave up my practice in the first years of my married life and occupied myself solely with chemistry and writing." So it was that he threw away his practice and settled down to translate works such as William Cullen's *A Treatise on the Materia Medica.*

It was during a reading of one of the treatise's chapters that Hahnemann found an explanation for why the bark of the cinchona tree (of which we now know contains quinine, a compound that is effective in killing malarial parasites) was a useful treatment for malaria. Hahnemann disagreed with Cullen's reasoning and on further contemplation found it curious that ingesting cinchona produced symptoms not unlike those of the very disease it was supposed to treat. It reminded Hahnemann of Hippocrates's "law of similars," which proclaimed that the cause of a disease might also help a patient regain their health.

The word "homeopathy" refers to the "like cures like" nature of the treatment's philosophy and remains a fundamental part of this alternative form of medicine. The mechanism behind its apparent success, Hahnemann maintained, was that the symptoms caused by a drug, being artificial and therefore weak replicas of the "real" disease, help the body's vital essence fight off the illness rather than directly contributing to it.

Diseases themselves were believed to be initially caused by "miasms" (not to be confused with miasma, which is defined as polluted or poisoned air), which were interruptions or "morbid derangement" of a person's vital essence. Left untreated, a miasm could progress from being something as simple as a cold or a skin rash into the body's core, making organs fail and eventually leading to death. One type of miasm, for example, was termed a "psora," which could begin with a skin condition such as scabies and eventually be responsible for various conditions such as cancer or epilepsy.

On finding a substance that could match a disease symptom, Hahnemann suggested a healthy subject needed to test it in order to prove its effect wasn't due to any disease. "Proving" in this manner over the years has led to a sizeable collection of remedies that work together to treat the numerous features of a disease. To treat a patient, a homeopath refers to both a repertory of disease symptoms and the *Materia Medica*—a compendium of remedies with their effects.

Of course, ingesting a substance that creates a set of symptoms only serves to make a sick patient feel even more ill. Hahnemann's answer was to mix the substance with a solvent and pound it ten times against the palm of the hand or another soft surface to extract its vital energy. Then, the solvent was diluted one part in 100, paradoxically making its effect on the body even more powerful. The dilution had the effect of softening the substance's impact on the patient's wellbeing. Needless to say, patients were far more appreciative of a sip of Hahnemann's bland tonic, than they were of being bled, sweated, given cramps by a physician or asked to drink a revolting mix of bitter herbs and alcohol.

Homeopathy relies on the solvent, whether it's water or alcohol, retaining information from a passing acquaintance with a solute. With each dilution, the original material is spread through more solvent, making it increasingly unlikely that any single tincture will contain so much as a single molecule of the active material. Consequentially, another feature of the solution must be responsible for its alleged effects. Traditional homeopaths explain this by referring to vague energies or essences, as if ghosts of the old molecules still haunt the solvent. Others have presumed there might be a more physical form of molecular imprinting taking place. French immunologist Jacques Benveniste published a paper in the journal Nature in 1988 describing the effects a solution of antibodies had on white blood cells called basophils, in spite of it having been diluted to homeopathic levels. This has led to speculation of solutes somehow changing the solvent's molecules in unique ways, creating a persistent system of chemical memory. Criticisms of Benveniste's method, coupled with the fact that few have been able to replicate his experiment, suggest that such a phenomenon doesn't exist.

If it did, the ramifications would be tremendous. Computing, chemical synthesis, geology, forensics, medicine . . . there isn't a field of science that would remain untouched by this single fact. Chemical traces would remain after the original atoms themselves had vanished, providing a link to past events. Information could be recorded as molecular substrates came and departed. The fact that the phenomenon is limited to observations that can also be accounted for by the psychology of the placebo effect suggests the alleged healing effects of homeopathy could simply be a matter of perception.

Given our understanding of chemistry, there are strong reasons why homeopathy should not work. Hahnemann's imaginative descriptions of miasms are contradicted by numerous experiments that lend weight to the role of micro-organisms, genetic variations or the result of malnutrition. More importantly, the presence of any undefined "essences" in either the patient or the remedy cannot be supported scientifically. There is no sign of any "ghost" that might possibly be left behind from a mix of molecules.

Hahnemann himself was well aware of the absurd reduction in the original material. Therefore, the important question is whether a solvent can retain a memory, spirit or essence of its previous constituents. While some modern experiments have alluded to the possibility, none have been repeatable under blinded conditions. Like Blondlot's N-rays, the mysterious effects claimed by so many vanish when investigated with the aid of the philosopher's toolbox, leaving us to suspect that such positive results have been caused by experimental error, rather than a real phenomenon.

Hahnemann's remedies were well received in his day, yet their popularity boomed after many became recognized in the United States as legitimate forms of medication in the *Food, Drug and Cosmetic Act* of

1938. Homeopathy continues to be promoted as a viable alternative to modern medicine, while being regarded by many as being more dangerous than beneficial given the propensity for people to avoid effective medication in its favor.

Homeopathy isn't alone, either. The term "complementary and alternative medicine" (CAM) is used to describe any form of treatment distinct from what is believed to be traditional medicine. What does this mean, exactly? Essentially, it refers to any medical intervention that has dubious scientific support or is directly contradicted by beliefs that have a greater weight of supporting scientific evidence.

Before the baby is thrown out with the bathwater, however, it's necessary to tease out the claims from alternative medicine's fundamental philosophy. Acupuncture is commonly recognized as a treatment that originated in China, and involves the careful and precise insertion of fine needles into various points over the surface of the body. Like homeopathy, the underlying mechanism relies on the existence of an essence or energy that defies scientific qualification.

Qi, as it is known, literally means breath or wind in Mandarin. However, the translation fails to capture the full meaning of the word, which also conveys a sense of animation or spirit that connects living things with the environment. Spending a day in the country, on a beach, feeling at one with nature can certainly provide a poetic sense of connection with the universe. Yet as strong as this sentiment might be, there is no single material law, energy or force that exists which subscribes precisely to the diverse properties understood as qi. It isn't just unquantifiable—it is virtually unqualifiable as well.

Investigations into whether acupuncture actually works have had somewhat mixed conclusions. A review of the practice by the World Health Organization (WHO) in 2003 resulted in a firm endorsement of the potential for acupuncture-based treatments. Unfortunately, the review was conducted on a large number of investigations performed over the years, consisting of a wide variety of sources.

This should be a good thing, right? Surely the more studies we have, the more confident we can be in the overall result. While large sample sizes can certainly be better than small ones, the nature of the studies themselves can have a substantial impact on the results that swamps any benefit. If highly criticized, poorly conducted investigations are thrown in with the more robust ones, a large sample won't be a good thing at all. Given that many of the trials included in the review originated in China, there is a good chance that, like Blondlot's French compatriots, national pride could be artificially inflating the confidence WHO has placed in the efficacy of acupuncture.

By comparison, a similar review by the independent Cochrane Collaboration concluded that the current evidence for acupuncture is either extremely weak or non-existent for most of its claims. A smattering of trials has suggested, however, that benefits in reducing the effects of pain and nausea can't be completely ruled out. Interestingly, while the needles themselves might elicit an effect, their location when placed on the body seems to be irrelevant. In other words, even if it does transpire that acupuncture has medical benefits, meridians and qi are superfluous complications.

Many CAMs attempt to distinguish themselves from traditional or "biophysical" medicine by claiming to be holistic, treating a person by taking into account more than just their physical condition. In fact, it is this approach that typically attracts many advocates. Gone are the days when your local general practitioner was available at short notice to chat about your cramps, bunions and relentless aches and pains. Complementary medicine seems more "real" on account of it appealing to our tribal brains, desperate for a kind word and a smile from somebody who wishes to hear more than just the symptoms.

When considering the culture of health and medicine, the functioning of the body is only half of the story. American medical anthropologist Allan Young suggests medical practice is itself a way of ascribing form and meaning to illness. He captualizods illness "as a kind of behavior which would be socially unacceptable (because it involves withdrawal or threatened withdrawal from customary responsibilities) if it were not that some means of exculpation is always provided." In understanding the philosophies fundamental to a medical practice it is necessary to begin with the motives behind not only those who are sick but other interested parties as well.

Disease exists within a tribal construct, as a difference between mere individual variation and social dysfunction. As such, viewing medicine only in terms of biology risks missing how treatment operates socially.

The family physician might once have been a vital part of the community, yet today long waiting lists and short visiting sessions have impacted an important piece of the health puzzle—the role of compassion. For many, illness is accompanied by a sense of alienation. A traditional homeopath will spend a good amount of time teasing out every symptom from the patient to craft their remedy. Acupuncture and reiki involve the warmth of another person's touch and the sense of personal connection with something greater than unemotional biochemistry and physics. The relationship between the practitioner and their client has an important role to play in regards to health and medicine.

In 2000, the American sociologist Erich Goode used select social characteristics of behavior to define groups within our broader community, with regards to paranormal belief. It was an attempt to classify the specific social spheres that kept the heart beating in various scientifically deceased concepts. One group he labeled the "client-practitioner," which could be identified by its tendency to employ pseudoscientific concepts in an exchange of goods or services. This group typically constitutes a mix of individuals whose views range between believing in their service without knowing it is scientifically unsound, believing in their service while being aware of the controversy or not believing in the validity of their service but relying on the controversy to dupe their clients into paying for something that does not work.

Ironically, the inability of science to support the claims of alternative medicine might be CAM's greatest advantage in gaining support. An absence or conflict in scientific evidence can easily be misread as a lack of true understanding, making room for mystery, and thereby possibility. Wherever there is the unknown, the imagination can run free and inflate hope to whatever limits a person desires. A homeopath might not truly be able to treat a child for colic, yet to the parent who is enduring the endless cries of a distressed infant, even false hope is preferable to the reality of suffering through it. Likewise, a person dying of an aggressive cancer that fails to respond to chemotherapy can still hold onto hope that a chance of survival lies in a field that science appears to know little about.

THE UNDEAD PRIEST

Teaching science in the United States state of Tennessee between 1925 and 1967 posed the risk of a hefty fine if care wasn't taken in how lessons were framed. On the passing of the state's *Butler Act* it became illegal "to teach any theory that denies the story of the Divine Creation of man as taught in the Bible." Asking your students to so much as open the chapter on evolution in their textbooks could see you stand before a jury, which was precisely what happened to high-school science teacher John Scopes on May 5, 1925.

Scopes vs The State of Tennessee concerned more than just a random science teacher falling afoul of a controversial law. Financed by the American Civil Liberties Union, the case was brought to test the state's enforcement of what was viewed to be an unjust bill. It was also encouraged by a group of Dayton businessmen who felt the public attention would benefit their town. Even though the defendant wasn't certain he had broken the law in practice, he was happy to stand before the court and admit his guilt in protest against the strict restriction. The proceedings that achieved fame as the "Scopes Monkey Trial" were never really about Scopes breaking the law; they were a judgment on the role of religion in the science classroom.

Given that a judiciary's role is to interpret the law and not rewrite it, Scopes never had a chance. He was sentenced to the modern equivalent of a $1,000 fine, overturned on appeal only because of a technicality in Tennessee law that gave the responsibility of invoking such hefty sentences to the jury and not the judge. The Butler Act would eventually be repealed after several decades of concern that other countries—namely communist states—were overtaking the United States in science education. Fear of communism trumped a compromise on religion in the end, and the Act was axed in 1967.

If it's difficult finding clear boundaries between "science" and "not science," finding an objective line between "religion" and "not religion" is even more problematic. Science writer Stephen Jay Gould attempts to resolve the conflict by claiming each field is capable only of answering questions relevant to its particular brand of philosophy. Termed "non-overlapping magisteria," or "NOMA," Gould contends that there is a clear delineation between the two authorities: science deals with fact and theory and therefore can't tell you whether it's right or wrong to abort your fetus; religion deals with morals and personal values, so can't tell you anything about the speed of light.

Gould's attempt at resolution implies religion, by definition, should have nothing to say about how the world was created, whether miracles are possible, whether a person's consciousness can exist independently of their brain or ultimately whether a powerful being can willingly modify the very laws science attempts to describe.

Would a religion be at all recognizable if these features were removed from it? Imagine Christianity without the resurrection of a person three days after death or Buddhism's cycle of *punarbhava*, or "rebecoming," without the reincarnation of immortal souls. Both describe beliefs on how the universe fundamentally operates; they describe laws that attempt to explain what we observe. Both are viewed as factual by many people of faith.

In reality, religion and science can't help but bleed into one another. Science steps on religion's toes when artifacts like the Shroud of Turin are carbon dated. Religion butts into science as the Bible is held up as proof that the world is only 6,000 years old, in direct contrast to observations of a slow-cooling sphere of radioactive rock. Two and a half thousand years of philosophical and scientific thought have provided humanity with a sense of faith in our ability to describe the universe and its laws with a degree of accuracy. Gods, angels, devils, ghosts, prayer, magic and miracles are no longer simple allegories or manifestations of our tribal brains, but are active components that must find a home in our physical universe or risk being made redundant. Modern science continues to struggle in its search for an adequate model explaining human consciousness, leaving many people to continue to embrace the inchoate concept of a soul or spirit that is capable of surviving death.

The supernatural is frequently invoked as an explanation for anything that initially leaves us scratching our heads. Abnormal psychology or neurology that we might associate with epilepsy, dementia and schizophrenia today was dismissed as the work of demons or evil spirits for millennia. Isaac Newton wasn't beyond suggesting the shortfalls in our theories could be a place where God resides. He felt that the orbits of the planets and comets would eventually pull and tug on one another, their gravity moving them out of orbit to disastrous ends. So God must gently prod them back into place every now and then. This claim prompted fellow physicist Gottfried Leibniz to say in a letter to a friend, "Sir Isaac Newton and his followers have also a very odd opinion concerning the work of God. According to their doctrine, God Almighty wants to wind up his watch from time to time; otherwise it would cease to move. He had not, it seems, sufficient foresight to make it a perpetual motion."

It is a habit of the tribal brain to apply supernatural explanations to gaps in our knowledge of the natural world. Our hunger for answers to our questions, coupled with our social engineering, satisfies our search for understanding with teleological, personalized explanations. In 2004, young Earth creationists on the Dover board of education in the US state of Pennsylvania even argued for a competition between alternative theories—evolution and a new model that proposed life was designed intelligently—claiming science could not explain questions on the origins of life and the nature of organic complexity.

The social landscape was vastly different from the one that prosecutors of John Scopes had to deal with, as a public awareness of science and civil liberties regarding separation of church and state had increased over the decades. There was no *Butler Act* equivalent and, therefore, no means of banning the teaching of human descent from non-human ancestors. Evolution was free to be taught.

The so-called theory of intelligent design was described as an alternative explanation for biodiversity and the origins of life. Fundamental to its claim was that modern theories of evolution were insufficient. The creationists referred to features described as "irreducibly complex," claiming there was no way a simpler version could function as anything. Key to it all was the belief that conscious intent was a necessary requirement for such complexity.

It seems fair to permit alternative theories to be taught in science. After all, science is only strengthened by competing ideas and arguments. And evolution is just another theory.

The problem with this perspective is that science does not operate this way. Theories might well be contingent and open to being abandoned or modified with new evidence, but that does not make all ideas equal. To compete with evolution, another theory would need to match its strength in explaining what we observe in nature. Intelligent design merely relies on the assumption that complexities that could not have resulted from mutation and selection must be the result of an intelligent creator.

The creationists' move exploited weaknesses in the public's understanding of science philosophy while appealing to their sense of fairness. They weren't asking that evolution not be taught; just that other theories had "equal time." What could be fairer than that?

In November 2004, the Dover Area School District publicly declared that teachers of year-nine biology at Dover High School would be required to read a statement arguing that intelligent design differs from "Darwin's view" and that the textbook *Of Pandas and People* would be offered as an alternative resource.

In 2005, parents of students who attended Dover High School sued the Dover Area School District over the board's ruling, resulting in a verdict that the forced reading of the statement was unconstitutional. The teaching of intelligent design as a scientific alternative to evolution was banned in science classrooms in Dover's school districts.

While this was a strong message revealing intelligent design to be little more than a thinly disguised copy of creationism, the trial inadvertently prompted evolutionists to take a closer look at the biological features intelligent design advocates had described as too complex to arise through evolution. Their shining example was the bacteria's flagellum—a wriggling tail, powered by a protein motor embedded in the microbe's wall, which propels it through a fluid. Microbiologists had offered no explanation on how the flagellum's integrated components had evolved together to form a tiny motor. Without any single piece, it would not function as a wriggling tail. As a result of the challenge, researchers subsequently determined that it had evolved from a structure serving a completely different function; a channel that pumped chemicals across the bacteria's wall.

Perhaps the most significant impediment religion imposes on science is the conflict over the importance of parsimony (Ockham's Razor) and symmetry (that the laws of nature are the same everywhere, at all times). While science maintains that the best way of describing what we see is to be sparing of our assumptions, religious philosophies have little problem with coloring a story with imaginative details or ignoring additional questions. While an intelligent creator would be an easy way of explaining how life originated on Earth, it presents even more troublesome questions. What is the nature of this intelligence? Why do the laws that govern our universe also have to possess intention and foresight? If complexity, such as the human brain, requires an intelligent creator, does the intelligence of the creator also demand its own creator? On top of that, science acts on the belief that the laws that govern our universe don't spontaneously change with time or space. In religion, there exist beings capable of perverting or acting outside of the very rules they are often deemed responsible for creating.

The best example of this conflict is in the need for one or more miracles to take place before a deceased person can qualify for sainthood in the Roman Catholic Church. In times past, a canon lawyer known as the Promoter of the Cause—or God's Advocate—would take reports of miracles associated with a blessed figure and argue their case for canonization. In opposition there would be a Promoter of the Faith—or Devil's Advocate—who would attempt to demonstrate that such miracles were fraudulent or mistakes, and do their best to show why they should not be promoted into sainthood. In 1983, Pope John Paul II reformed the process by merging the roles into a single figure, known as the Promoter of Justice. During his papacy the number of canonizations soared to nearly 500, making him the most saint-happy pontiff in history (compared with a total of 98 between the beginning of the 20th century and the beginning of his rule).

By definition, a miracle is an act of intervention in the expected course of nature by a deity. In the case of Mary MacKillop's canonization, in December 2009 the Congregation of the Causes of Saints issued a papal decree that the recovery of an Australian woman who had been diagnosed with a primary cancer of the lungs and secondary cancer of the brain was due to God's direct involvement in changing the normal course of the disease. It would be a contradiction to presume such a fortunate turn in events operated outside of the laws of biology in a scientifically explicable fashion, and an unnecessary assumption that it involved the actions of a powerful intelligent entity.

A belief in the supernatural creation of the universe is debatably well outside the scope of scientific inquiry itself. On one hand, it introduces an unnecessary layer of complexity without explaining any-thing—it might be convenient to have God flick the switch to make the machine whir to life, but now there is that annoying question of how God came to. Yet if a divine universe looks identical to one where God is absent, it cannot be shown to be impossible, even if it is superfluous.

Given science is an attempt to define observations of the natural, giving credence to the supernatural is a contradictory concept. For some, science is not a matter of getting to know how their god might have created the universe, but of gilding the lily of their faith. Healthy hands and feet spontaneously bleed in sympathy with Christ's cruci-fixion; holy statues weep sweet, salty or oily tears; apparitions mani-fest; prayers are heard; tumors vanish . . . all taken as empirical evi-dence of a divine presence in the here and now.

People gather in the belief that science has somehow made room for their faith, either by turning the other way while a miracle takes place or by permitting a special set of rules just for their choice of divine superpower.

In addition to his category of the client-practitioner, Erich Goode describes a collective based on religious faith that includes all beliefs involving supernatural forces. Where the client-practitioner is a one-way economy of pseudoscience and paranormal concepts being sold to the public, the social context of religion relies on receiving personal strength through a communal, two-way sharing of a paranormal idea.

Could faith in a personalized universe be a consequence of a brain that evolved to negotiate with other personalities rather than just deal with facts and figures? A belief in the controlling hand of the supernat-ural might lend a sense of purpose to an otherwise chaotic and hostile universe, while the absolute certainty that comes with faith in a scrip-ture or in the words of a religious authority is often far more palatable than the necessary doubt and aggressive criticism inherent in science. A brain built to see the world in social, personal tones will naturally tend toward the question "why?" as if a creative agent imposed a pur-pose on its creation.

In some ways, God might be far more at home in our brain than sci-ence could ever be.

THE UNDEAD MIND

Of all scientific disciplines, those that involve studying the mind are the most troublesome. The sheer complexity of the brain's functioning makes it all but impossible to come up with simple, universal rules that govern all of an individual's behaviors; not to mention the need to take ethics into consideration when dealing with human subjects limits options in coming up with experiments. Psychology suffers from the frustration of being one of the most personally useful subjects for science to explore and yet the one that we have the most difficulty teasing apart.

Of all the myths to do with the brain, one of the most pervasive throughout the last century has regarded the potential for untapped capacity. A commonly repeated article of baseless trivia is the claim that we use only 10 percent of our brain, which appeals to believers in psychic talents, such as Uri Geller who states in his *Little Book of Mind Power*, "Our Mind-Power is like an iceberg, with 90 percent lying hidden beneath the surface."

It's not entirely clear where this figure came from initially, whether it was a miscommunication based on early estimates of the number of neurons in the brain compared with their supporting glial cells or a direct quote on latent mental abilities. Either way, while the brain is amazingly plastic it is also wonderfully efficient, so none of it goes to waste. There is not a single area of the brain that can sustain damage and not have some impact on our behavior, even if the brain is capable of overcoming some forms of impairment through "rewiring" neurological pathways. Through imaging tools we can actually see the brain use energy as it operates, watching it tick. While some parts will be quieter than others, at some point all of the brain's circuits are devoted toward accomplishing a particular task.

The notion of unleashing hidden skills is alluring. We take for granted how our brain performs its tasks such that aberrations can seem almost mystical. With impediments in some regions, those with conditions such as savantism can demonstrate extraordinary prowess in others. The American Kim Peek, famed as the inspiration for the 1988 movie *Rain Man*, was reported to be able to read two pages at the same time, covering each in approximately ten seconds, and could recall text from 12,000 books. Yet he could not walk until the age of four, had poor motor skills and demonstrated low scores on intelligence tests. Neurologically he had been born without a corpus callosum. While it's not clear how these traits are all related, there persists the temptation to consider the possibility that we're all capable of improving how our brains fundamentally operate.

The organization Brain Gym International was formed in 1987 as the Educational Kinesiology Foundation, changing its name in 2000. It describes on its website how a variety of bodily motions mimic similar movements performed during the first years of life as we develop coordination. The program claims to improve psychological attributes such as memory, concentration, academic skills and self-responsibility. Naturally, the program is aimed at teachers and other educators as a way to improve learning in their students.

Educational kinesiology is an offshoot of a practice called "applied kinesiology" (AK), a system developed in the 1960s as a mix of chiropractic and traditional Chinese medicine. Practically speaking, it can be described as using the perception of changes in muscle tension to diagnose medical problems, which can then be treated with adjustments to the spine. Proponents often go so far as to believe that the body "knows" what is good or bad for it even if the mind doesn't, as measured by testing the resistance of the body's muscles to an external strain.

Unfortunately, these effects typically vanish when performed under randomized, double blinded conditions, implying that AK as a diagnostic tool seems to be no more useful than guessing.

Brain Gym International's claims suffer from the same problem. A document on their website cites numerous studies that support its claims. However, on closer inspection most of them are reference studies published in their own journal—hardly the adequate criticism from a diverse field of researchers one might expect. Other studies are tenuous and show no sign of having been replicated by other institutions. Even Brain Gym International concedes on their website that most of their evidence is anecdotal, meaning it relies on the feedback from teachers who have used the method and believe it works.

An example exercise suggested in the Brain Gym teaching manual includes pressing the tongue against the roof of the mouth. Along with other exercises where fingers are linked in order to "connect the electrical circuits in the body," this exercise is said to stimulate the limbic system, which is responsible for emotions, and encourage it to act "in concert" with the frontal lobes, where reasoning takes place. None of this makes any sense from a neurological perspective, even though the language used to describe it sounds scientific.

While a proportion of the program seems to rely substantially on pseudoscience, many of its claims rely on good pedagogical practices. Goal setting, focusing on preparing to learn, integrating learning into physical movements and so forth are practices any experienced teacher should be familiar with. Drinking water before a test is indeed a good suggestion, given that dehydration can affect our ability to think clearly. It's common to find good science mixed in with pseudoscientific fields, making it difficult to tar the entire concept with the same brush.

Exercise and movement are certainly important components of any educational program, yet it is the claimed mechanisms behind the practices advocated by the Brain Gym program that aren't scientifically supported. As with any pseudoscience, the language describing educational kinesiology might make it sound robust and trustworthy, but that confidence isn't matched by proper application of the philosopher's toolbox.

Some conditions, such as extreme autism or cerebral palsy, dramatically impact on a person's ability to communicate their needs to the outside world. In 1977, Rosemary Crossley was a teacher at Victoria's now-closed institution for severely handicapped children, St Nicholas Hospital. To the astonishment of both the hospital and the Health Commission of Victoria, she argued that 12 students in her charge who had been diagnosed with conditions such as cerebral palsy had not only succeeded in communicating with her, but were demonstrating normal intelligence.

By all accounts, Rosemary was a devoted, caring teacher of children whose needs went far beyond that of the average student. By modern standards, St Nicholas was a rather hellish place for any child to find themselves in, let alone one whose wellbeing and survival relied on the services of unaffectionate and untrained nurses and aloof, stoic doctors. Rosemary's affection for those children earned her the respect of many, even while her claims were scrutinized and dismissed by others.

Rosemary's techniques for communicating with her students involved assisting their movements in order for them to indicate which words or letters they wished to articulate. This is most typically done through carefully guiding the student's hand over a selection of pictures or a keyboard of letters and numerals and noting the letter or image that coincides with a sign from the subject.

Following an investigation of Rosemary's work in 1989, the American sociologist Douglas Biklen was so convinced by its success he took the practice back to the United States and established the Facilitated Communication Institute at Syracuse University. In the years that followed, the method spread across America and then the world. Parents of severely disabled children, desperate to gain an insight into their child's personality and wellbeing, embraced this radical new communications tool.

At the same time, some rather critical questions were being asked. Why were children using words that would normally be considered well beyond their years and experience? Worse still, some facilitators noticed that answers were being provided in spite of the child's distracted focus. Was it possible that the facilitator was truly responsible for some, if not all, of the communication?

Acting as a medium for the disabled might be considered relatively harmless, especially if the responses are of comfort to the parents and relatives, yet on occasions where the message has serious repercussions, it's important to know whether the results are mostly the intentions of the subject or the beliefs of the facilitator. Where questions of abuse arise as a result of miscommunication, people's lives can be destroyed unnecessarily should the accusation be baseless.

Today, the balance of evidence weighs heavily against facilitated communication. The American Psychological Association has declared that it is unproved with "no scientifically demonstrated support for its efficacy." On the other hand, the Autism National Committee feels the practice shouldn't be dismissed on the back of a few "flawed studies that are poorly designed and/or whose results are incorrectly extrapolated to the entire population of FC users." Proponents argue that the results do indeed falter under controlled conditions, as the clinical nature of the testing confounds and confuses the subjects.

If there is something to facilitated communication that goes beyond the unconscious control of the facilitator, it is clearly difficult to distinguish. Indeed, individuals with severe physical disabilities aren't always encumbered with an impaired brain. One need look only to the British physicist Stephen Hawking to see how a communications tool can help a brilliant mind bypass an uncooperative body. Yet there can be little doubt that the words spoken by Hawking's computer started as thoughts in his head. The same cannot be said for those who are assisted by the guiding hand of another mind that is subject to both bias and the troublesome "ideomotor" effect.

We function under the illusion that most of our muscles' movements are conscious and intentional, especially those that aren't related to the heart or the digestive system. Most are; however, to achieve this our brain performs some rather complex tasks that involve a great deal of information filtering. In cases where this filtering equipment falters we experience debilitating conditions like Parkinson's disease. Yet even in otherwise healthy brains, tiny commands seep through, nudging muscles in near-imperceptible ways. This ideomotor phenomenon is sufficient to explain all manner of other spooky effects, from Ouija boards to the swinging of a clairvoyant's pendulum, without invoking supernatural forces.

Authorities in countries such as the US and Canada employ the use of lie detectors to identify guilt in criminal suspects or even employee screenings, even if the reliability of this practice is highly questionable and often inadmissible as evidence in court. Torture continues to be used to coerce captives into revealing what they know, in spite of the high risk of false or useless information being fabricated as a direct result of the process. Racial profiles are applied to a crowd of passengers filing through airport security in the mistaken belief that it might help increase the chance of catching a terrorist, even if the end results waste resources and potentially distract law enforcement from preventing or solving an actual crime. Even the humble horoscope in the back of the daily newspaper is given undue credibility in its attempt to describe the minds, hearts and souls of its readers.

Science as a search for knowledge consists of overlapping fields of research that rest upon numerous fundamental theories and laws. The basic laws used by a nuclear physicist to make sense of the forces within the center of an atom are mostly the same as those an astronomer uses to make sense of the various life-cycles of different stars, for instance. Disciplines cannot operate in isolation from one another.

Goode's social category of the classic pseudoscientist, however, casts them as pariahs with respect to related fields of research. Professional astrologers won't have a great deal in common with astronomers and are unlikely to present talks alongside representatives from NASA. Those who conclude telekinesis is real won't typically find themselves presenting those findings at neurology conferences. In psychology, concepts that lose favor or fail to gain appeal among the profession tend to persist external to the field, discussed in exclusive journals that are edited favorably or often with less rigor.

In defense, members of Goode's pseudoscientist social group will often protest that their ideas aren't wrong, but are simply ahead of their time or yet to be proven; they envision a future where concepts such as remote viewing, extrasensory perception, telekinesis or applied kinesiology are no longer dismissed or viewed as unscientific.

As science progresses to tease out the mysteries of the brain and understand what makes it tick, we will undoubtedly place too much or too little confidence in much of what we learn. After all, we all have a mind, so we all believe we know quite a bit about it. Whether we truly do or not is another matter.

THE UNDEAD BEAST

Erik Ludvigsen Pontoppidan, the 18th-century Bishop of Bergen, would have been rather disappointed with the relatively tiny size of the tentacled beast hooked by New Zealand fishermen in Antarctica's Ross Sea in February 2007. It weighed a paltry 495 kilograms and was a mere 10 meters from the tip of its longest tentacles to the point of its mantle. When alive, its eye might have been only 40 centimeters in diameter at the most.

The bishop might have considered this particular colossal squid a poor match for the Kraken—a legendary sea monster he claimed in his book *Natural History of Norway* to be virtually a floating island at one and a half miles wide (nearly two and a half kilometers) with arms like a starfish—but to cryptozoologists, the proven existence of *Mesony-*

choteuthis hamiltoni is evidence that allegedly mythological creatures can be found lurking in the globe's unexplored shadows.

New species of organism pop up all the time in biology. Some are right under our noses, mistakenly ignored until somebody identifies a key difference that separates it from its cousins. Others are discovered as new terrain is explored and scientists are able to record and categorize the flora and fauna they stumble across.

While new species aren't uncommon, uncovering a completely new class, or even family, would be the find of a lifetime for any biologist. In 2006, a new family of hairy-looking crabs was found in the ocean off Easter Island by an international team of marine scientists, indicating that the unknown depths of the oceans are a frontier full of promise for any zoologist eager to claim naming rights over a new crustacean or mollusk.

Cryptozoology is a topic discussed in whispers as if it is zoology's embarrassing stepson. Yet like much else in science, it's hard to know exactly when a person crosses the line from useful, genuine research in the strange and unusual to engaging in pseudoscience. At one extreme, disciplinarians in this field might express an interest in searching for aliens, fairies, dragons or demonic el chupacabras (Latin American "goat suckers"). At the other end, researchers spend their hours looking for signs of animals that are out of place, such as panthers or cougars on the Scottish moors, or out of time, like the recently extinct Tasmanian tiger. In between there are beliefs in strange lake monsters, seven-foot-tall ape-men and modern populations of dinosaurs.

"Cryptids," as such mysterious critters are termed, often enter folklore as a rumor, hoax or legend that resists efforts to be substantiated with an appropriate level of hard evidence. They persist on the backs of reported sightings or enthusiastic interpretations of historical documents, paintings or oral traditions.

Old Scottish legends tell of horse-like river demons called "kelpies" and their lake brethren, the each uisge (pronounced "ack ice-ah"), which drowned children eager to jump on their backs for a ride. One of the oldest accounts of a strange animal living in the waters connected to Loch Ness involved a 6th-century Irish monk who was later canonized as Saint Columba. While standing on the shore of the River Ness, he reportedly commanded a so-called water beast to keep away from his companion who was taking a swim in its chilly waters.

Although it's commonly told as the oldest account of a Loch Ness Monster sighting, the theme of saints holding sway over wild beasts of the land and water as proof of their divinity makes it hard to separate from allegory. Was it an unearthly monster, a spiritual symbol or just some undescribed animal? Given no other information, we can only guess at how much is metaphor, how much is myth and how much is accurate observation.

It wasn't until July 1933 that the monster's modern incarnation developed. A gentleman by the name of George Spicer reported to the local paper, the *Inverness Courier*, that he and his wife had just seen a long-necked creature 25 feet (7.6 meters) in length traverse the road on which they were traveler, with a lamb in its mouth. This description was the first of what would become not only "Nessie's" unmistakable profile, but also that of lake monsters everywhere; a snake-like neck, bulky torso, short legs and a humped back.

Several months prior to Mr Spicer's historic sighting, the movie *King Kong* had debuted in Hollywood and quickly captured the attention of audiences everywhere with its stop-motion animation of dinosaurs and a giant gorilla. Legends of horse-like demons might have been common around the Scottish waterways, but there are no recordings of hump-backed serpentine creatures gracing the shores of any loch before 1933. Is it possible that George's imagination took inspiration from the sauropods or plesiosaurs he might have seen in the newly released blockbuster?

In the decades since, there has been no shortage of eager witnesses who claim to have seen the monster either in the loch or lumbering along its banks. There has also been a substantial amount of hoaxed evidence, such as footprints pressed into the soil and hazy photographs of scale monster models. More recent ultrasound images taken in 1972 by the American lawyer and avid "Nessie" researcher Robert Rines are said to show a plesiosaur-like flipper under the water. While one of the retouched pictures published by the press could very well show a fin mid-sweep, the initial computer-enhanced image is less clear. Later attempts by third parties to replicate the image from the original data failed.

Loch Ness is hardly an isolated valley in the middle of nowhere, so how could a beast larger than an elephant live in such a populated area for so long and only attract attention in the early 1930s? Was it new to the loch? While the area has been populated for centuries, it's often argued that the sheer depth of the ravine could be enough to explain the creature's elusive nature. Yet for a sustained population of massive animals to survive long periods in the murky depths of Loch Ness, it would require a number of unique features that do not resemble any category of known organism.

To exist, a large number of questions would demand some robust answers. For instance, why are there no bodily remains? How could a population be sustained on the loch's thin food supply? Why are there no historical accounts of long-necked fauna in the region? Speculation is rife, but the complexity required of most answers makes the existence of such animals less likely when compared with the possibility that the accounts are a generous collection of hoaxes and mistakes.

The Loch Ness Monster's pathological shyness is a feature common to many cryptids. Whether it's the oversized hominin Bigfoot or the last of the Tasmanian tigers, proof is always limited to articles that can be easily faked, such as footprints or out-of-focus film. For many hunters of strange beasts, fuzzy photographs and oversized plaster casts provide enough incentive to keep looking. Yet their confidence fails to be impeded by factors that should otherwise limit their enthusiasm, moving them toward the barren fields of pseudoscience.

Cryptozoologists who maintain a belief in their favorite cryptid's existence will tend to point out instances where other creatures presumed to be long extinct or flights of fancy have indeed been discovered. The coelacanth, for instance, is exemplified as a "living fossil" related to lungfish, long presumed to be extinct prior to being rediscovered in South Africa in 1938. Mammals such as the okapi in central Africa and the mountain gorilla were presumed to be just local legends until hunters brought back first their remains and then living specimens.

In each case, skepticism was tempered by limited knowledge of the world and its environments. As exploration increased, those troublesome questions were met with solid answers, until eventually bones, pelts and then live specimens were revealed to the world, expanding not just our knowledge of what was real, but creating a scaffold for what was possible. Coelacanths might have belonged to a rather exclusive club, but their unlikely existence was quickly substantiated when researchers made an effort to look closer.

Yet a sighting of the Tasmanian tiger, extinct since the 1930s, might raise questions that demand more complicated answers. Justifying a search for a *thylacine* would be more difficult than it would for a more recently extinct organism, given that hoax or mistake has an increasingly higher chance of explaining the sighting than the ghost-like existence of an animal that has left no trace for decades. With every fruitless search there are more questions that would require increasingly detailed answers, further reducing the probability of success.

Cryptozoologists come from all fields of interest, experience and expertise. Some are researchers with years of academic experience in a specific scientific field. Others include seasoned outdoors-men, amateur hunters, adolescents interested in the unknown and the elderly who once thought they caught a glimpse of something unusual. In addition to the pseudoscientist, the religious and the client-practitioner social groups, Goode described the "grassroots" community of believers. All across the world there are groups of average men and women with typical day jobs who meet periodically to discuss or even search for signs of mysterious creatures such as Bigfoot. Their beliefs are maintained by the regular interactions with others who regale the gathering with tales of close encounters and near misses, retelling famous anecdotes and legends that set the imagination alight with possibility. While Goode's category of pseudoscientist congregates at conferences to present papers and publish their results in journals, the grassroots believer gathers at conventions to engage in less formal talks, publishing their opinions in magazines, books and newsletters.

The promise of a world-changing discovery is what drives many on in their search, along with personal curiosity, the possibilities of fame, acclaim and fortune, and the pride that comes with solving a problem that stumped others. In most cases a keen interest in the strange and unlikely existence of certain animals is harmless. But when hope replaces reason in the search for new species, and resources are diverted to a search that is almost certainly going to come up empty-handed, what starts as good science can quickly cause people to gain confidence in ideas that have little chance of proving productive.

SILVER BULLETS

Under the palpating fingers of a physician at the beginning of the 20th century, you might have been diagnosed with ptosis of an organ, where the root cause of your suffering would be blamed on the slippage of a kidney or a loose uterus. It was an era of anatomy when the body's viscera were considered fixed in place, and any deviation was viewed as pathological. Treatment was simple—complete removal of the offending organ.

Drooping organs are no longer useful medical concepts, abandoned as improvements in surgery and anatomical exploration through the early 20th century demonstrated no link between the displacement of kidneys or ovaries and symptoms. Which is just as well—the price of the idea was unnecessary pain and suffering on the surgery table, debilitation with the loss of a healthy organ and, in the worst cases, the loss of life. It was with great fortune that the belief lost favor among surgeons.

While some ideas pass away silently with the last of their advocates, others persist through the generations, passed down within tribes who maintain the faith in the face of skepticism. Unfortunately, there is no silver bullet that can erase an idea forever. Resistance can form in the protests of dissenters or vocal opposition ridiculing the deluded, but with conflicting epistemologies vying to be authorities of information, the battle to challenge perceived myths and untruths requires more than an opposing opinion.

Chapter 7

THE TANGLED WEB

Who is in control of what we know?

Ａll human tribes invariably operate within a fluid hierarchy of authority that changes with promotion, failure, death and success. Our ancestors groomed those whom they respected, feared or admired, and in turn were groomed by those who respected, feared or admired them. While conversation has taken the place of lice harvesting, we continue to respond to the social tides of obedience and subservience. Even within the most democratic tribes there will be those who exercise their expression freely and those who struggle to be heard; those whose opinions are given free rein and those who are forced to argue.

WHEN WORLDS COLLIDE

It is said that the Renaissance astronomer Nicolaus Copernicus died on the very day he finally held the first printed copy of his book *On the Revolutions of the Celestial Spheres*. His influential work on planetary motion had developed from a simple manuscript titled "Little Commentary," which listed seven basic assumptions about the movements of celestial objects. Following nearly two decades of collecting detailed accounts of the shifting heavens, the book was more or less complete. However, Copernicus resisted publishing it for a further decade, taking his time in tinkering with it and seeking the opinions of his close friends and pupils. The astronomer was sensitive to the potential impact his revolutionary theory might have on the world, fearing how others would react to the thought that the Earth turned on its axis and slowly moved around a stationary sun.

For most of history, people thought that the planet beneath our feet was immobile. Instead, it was the heavenly bodies that crept steadily along their distant paths on immense vaults that turned over our heads. There were occasional flickers of imaginative speculation, such as that of the 4th-century BCE Pythagorean philosopher Philolaus, who proposed a model of the universe where all objects—including the sun—orbited a central fire. Yet by the 15th century few had dared to challenge the speculations of the Roman philosopher Claudius Ptolemaeus, who in the 2nd century CE described the universe as a number of concentric spheres, with the Earth at its core.

Scale and inertia compromise any direct sense of the Earth's movement, making Ptolemaeus's "geocentric" theory feel intuitively correct. Yet a sticking point that had persisted through the centuries was the astronomical phenomenon called "retrograde," where a planet's journey as seen from Earth appears to slow, pause and reverse for a short distance before moving forward again. The sun and the moon, on the other hand, don't behave this way. Ptolemaeus's system could not adequately explain this strange S-shaped path of some of the celestial spheres.

Copernicus reasoned that if the Earth and the planets circled the sun, their relative motion would affect how an observer traced their velocity. Variations in the sizes of the orbits would mean some planets made it around the sun in less time than others; therefore, when standing on a moving Earth, a planet would appear to slow down as it approached, reverse as the Earth passed and return to a forward motion as our orbit took us around the sun. He believed the moon was the only body that did circle the Earth, which explained why it didn't exhibit retrograde.

There's no clear indication of what concerned Copernicus more—scorn from religious authorities or a backlash from his astronomical colleagues over challenging such an established view. Either way, its receipt in the years after his passing failed to elicit the controversy he'd expected. A single Dominican theologian was the first to denounce the theory in an appendix to his manuscript on the scriptures, and even that was a full three years after Revolutions had been published. Officially, the religious authorities had little to say on the matter until early the following century, when an Italian astronomer faced a decision regarding its defense.

Galileo Galilei's early career was as a mathematician, tutoring in subjects such as geometry, mechanics and astronomy at the university at Pisa. He considered Copernicus's heliocentric theory of planetary motion to be superior to the old geocentric model, mistakenly believing that the ebb and flow of the ocean's tides were caused by variations in the speed of rotation of the Earth, causing the water to slosh around. Johannes Kepler—a contemporary astronomer in Germany—had promoted the view that the moon's pull had something to do with tides instead, which was flatly rejected by Galileo as pure nonsense. He was confident that the Earth turned and the tides were proof.

Nonetheless, although he would later be proven incorrect on the cause of tides, subsequent discoveries further cemented his opinion and would come to be the evidence required to change the established order. In 1608, a Dutch lens maker produced what has been credited as the world's first telescope. News of the innovation spread, and Galileo set out to construct his own in order to get a good look at the night sky. It was with this novel instrument he discovered that Jupiter was flanked by a line of smaller, dimmer stars. More importantly, as he gazed at this family of objects night after night he noticed they gradually changed their position, as if they were planets in orbit around Jupiter. Here was a direct observation of wandering stars that circled something other than Earth.

Later that same year Galileo turned his attention to Venus. As the months passed he watched it gradually wax and wane in shape and brightness, in a fashion similar to the lunar phases. Like the retrograde movements of the planets, it was easier to justify these observations according to Copernicus's model than a geocentric one.

It was in 1613, at a dinner party attended by a former student of Galileo's, a philosopher, Galileo's patron and the patron's mother, that the seeds of the astronomer's troubles would be planted. A discussion on his discoveries ensued, where the patron's mother asked the former student to show why she should accept such a contrary account of the universe over the words of the Bible. The scripture was clear on the Earth's static nature, stating in Psalms 104:5 it "can never be moved." There are numerous other instances of the sun traversing the sky—and even being forced to pause in its journey for an entire day—implying it must move while the Earth does not.

On hearing of this conversation, Galileo penned a letter which gave his view on the subject, which came to be known as the "Letter to the Grand Duchess Christina." It was this letter that eventually found its way into the hands of the Catholic Church's infamous Roman Inquisition, a tribunal system of the Holy See established in the early 17th century to prosecute individuals accused of heresy.

In 1615, a respected theologian and cardinal by the name of Roberto Bellarmino received a manuscript written by a Carmelite priest attempting to reconcile Copernicus's writings with the scriptures. Bellarmino realized he couldn't simply dismiss an astronomer's factual observations out of hand. Yet such notions bordered on heresy, contradicting the clear, unambiguous claims of the sacred scripture. As a solution he made a call for direct, physical proof that Copernicus had the right idea and asked that his book not be banned outright, but instead advised it should be removed from circulation and modified to reflect this need for definitive evidence. His seemingly tolerant suggestion opened the way for the heliocentric model to be taught as an interesting idea, and even as a mathematical tool for making predictions, but it was by no means to be communicated as if it was a physical reality.

Bellarmino's call for proof was a Sisyphean challenge, making it impossible for Galileo to satisfy. It was a safe bet that the requirement for suitable absolute proof could not be met, therefore keeping the scripture's model of the universe safe without seeming as if he was being deliberately close minded against a competing theory.

The Biblical text might have been a literal account to Bellarmino, but Galileo looked to the advice of the 12th-century Christian philosopher Thomas Aquinas, who suggested in his work *Summa Theologica* that the scriptures contained levels of meaning that require all of the body's senses to interpret. Galileo believed that there was no conflict between science and the faith and that it was possible to reconcile his observations with the conflicting passages. But under threat of being accused of heresy by the Inquisition, he was driven to take the fight to Rome and personally defend his case.

His dilemma was rather straightforward—on one hand, he could simply maintain that his opinion was purely hypothetical. Based on Bellarmino's leniency, he would be permitted to advocate the heliocentric model so long as it wasn't presented as a physical truth. However, Bellarmino justified his concession on the grounds that the theory had no convincing physical evidence to support it, which Galileo simply could not agree to given what he had observed with his telescope.

The case was put before the Inquisition's tribunal, who concluded that since it contradicted passages in the Bible it was plainly a heretical belief. The ruling led to a prohibition on the publishing of any work that supported Copernicus's theory of heliocentrism, including that of the Carmelite priest's. To press his case further, Galileo risked incarceration and ecclesiastical condemnation, if not worse.

His meeting with church officials, including an audience with Pope Paul V, was a surprisingly congenial one. Galileo begrudgingly relented in his cause, agreeing that he would neither "hold nor defend" Copernicus's theory as truth. While it technically permitted him to communicate the theory, any attempts at teaching it would require precise wording to avoid censorship by the Church.

As a commodity, the value of knowledge can purchase freedoms, while its theft or corruption can lead people to defend it with great passion. In 17th-century Europe, absolute knowledge as defined by a sacred scripture was of greater use to those with wealth and political influence than the fragile musings of natural philosophers. The authority of the Catholic Church identified a threat in the inconsistencies between observed reality and scriptural dictations and saw fit to use intimidation and aggression to continue to reinforce a view that was starting to fray at the edges.

Bellarmino's solution in diverse realities—a mathematically useful one that describes observations and a contrasting absolute, dogmatic truth—is a clear reflection of humanity's comfort in mythopoeism, where inconsistencies between accounts are overlooked or reconciled rather than viewed as mutually exclusive. Faced with anomalies, our brain is capable of justifying them to retain the beliefs of our tribe. Even Galileo felt obliged to compartmentalize the differences between what he saw and his society's spiritual dogma, categorizing his scientific conclusions by way of facts and rationalizing scriptural accounts as metaphorical truths. To not do this would jeopardize his position in the wider community and challenge the values he'd been raised to appreciate.

Indeed, it was thanks to his relationships with those in power that Galileo found a way to publish a work on the topic of planetary motion. On receiving the papacy in 1623, Pope Urban VIII looked favorably on the astronomer, and conceded to the book's publishing on several conditions, such as that the book had to provide arguments for and against both models and could not openly favor heliocentrism.

In 1632, Galileo's book *Dialogue Concerning the Two Chief World Systems* made its way into the public's hands, discussing the topic of heliocentrism indirectly through the words of a trio of characters—a Copernican astronomer, a philosopher and an Aristotelian—who were engaged in an argument. It was this final character, a man named "Simplicio," who gave voice to the traditional view of geocentrism. By cleverly making Simplicio to appear foolish and prone to emotional outbursts, Galileo made his opinion clear without resorting to explicit statements of fact. Unfortunately, the satirical angle and thinly veiled mockery was the book's undoing. Urban had stipulated that his opinions were also to be included in the book, which naturally could only be articulated by the Aristotelian. Regardless of whether or not Simplicio was intentionally modeled on the pope, the Church's cardinals insisted it was so and pressed Urban to respond forcefully. The friendship that had led to the book's publishing had been pushed to its limits and Galileo was sent to trial for heresy.

A panel of theologians judged *Dialogues* against Bellarmino's ruling that Galileo should refrain from teaching heliocentrism as if it was a universal truth, and found he had compromised the initial decree. His book was banned and he was sentenced to imprisonment, which was subsequently reduced to house arrest for the remainder of his life.

Today, Galileo is popularly regarded as a founding patron of modern science. His discoveries have resonated through the ages and eventually tipped the balance in favor of the heliocentric model of planetary motion. As with any good theory, the model has continued to change over the centuries. No longer can we view the sun as the center of the universe—indeed, the nature of the universe's shape and expanding structure has made it difficult to state unequivocally that such a thing even exists. However, without such contributions as those made by Galileo in the early 17th century, it would have been impossible to make such progress.

The flow and evolution of our collective knowledge is dependent on its distribution. Controlled by the scope and values of a minority, it is deprived of the diverse minds of other tribes, who in turn hold knowledge up to their perceptions of reality and judge it accordingly. Fortunately, the authorities of the western world were replaced with time, while Galileo's observations remained for anybody to see. Jupiter's moons and Venus's phases were just as true a century later as they had been during the 17th century; the authority of the Church, on the other hand, faded with the progress of the era referred to as the Enlightenment.

POWER TO THE PEOPLE

In the aphorism *scientia potentia est*, Francis Bacon famously described knowledge as power. Given that knowledge is a resource that perpetuates itself by being shared between people, those who are in a position to communicate it are the ones who are truly in charge. Do those in a position of influence over a community's wellbeing have a responsibility to influence what it believes, evaluating the knowledge it receives and censoring that which they identify as potentially damaging? Or should all information be promoted freely regardless of its apparent worth, open for individuals to personally discern as useful or fabricated nonsense?

These are difficult questions to answer. Evolutionists are often reluctant to debate creationists for fear of inadvertently lending credibility to what they see as a baseless argument. Intelligent design advocates have called for equal time in the classroom in the name of fairness. Should Holocaust deniers have the same opportunity to present their views as conventional Second World War documenters? What about permitting anti-vaccination groups to air their opinion on the potential dangers of vaccination? Should instructions for making explosives be freely available? Euthanasia instructions? When, if ever, should an authority dictate or prevent the dissemination of information?

Of course, various authorities make such decisions all the time, controlling the context of the knowledge they are responsible for communicating. Schools determine what will be covered by curriculum and what will be left out. Newspapers give space to stories that they deem to be interesting and "newsworthy." Even the most democratic governments will have laws to control how products can be advertised and to govern the dissemination of views that might inspire hatred or violence. How knowledge is framed can change how it is received, robbing it of necessary meaning and acting as a form of censorship. Information is never free of context, even when it is presented in an apparently naked state, making it hard for any single authority to ever be a truly free source of knowledge.

In November 2009, a string of e-mails was hacked from a computer server at the University of East Anglia's Climate Research Unit, and leaked to the media. Dubbed "Climategate," it brought to the public's attention an exchange of personal messages between climatologists on how they should deal with anomalous details from a recent study on the effect of historical temperatures on tree growth. Terms like "trick" and "hide the trend" hinted at nefarious intentions on the part of the climatologists—a boon for skeptics who felt climate change was a work of fiction created by scientists desperate for funds.

For the most part, the e-mails described nothing so sinister. Investigations into the matter failed to find any evidence that there had been an attempt to censor data in an effort to subvert the peer review process, although a review panel chaired by a British senior civil servant did describe the scientists as "unhelpful and defensive" in response to legitimate freedom of information submissions. Some petty squabbling and an expression of mistrust of another scientist's intentions were as conspiratorial as it got; in the context of the climatologists' research, "hiding the trend" had nothing to do with deceit, but related to sorting the relevant results from the irrelevant ones. Yet a small amount of knowledge robbed of its contextual framing could be easily presented as a convoluted conspiracy where climatologists were considered to have engaged in a scientific swindle.

It's hard to put a precise value on the public's trust in the role science plays in the decisions made by authorities. In its ideal form, science is the most democratic way of determining a course of action, considered as free of bias and personal agenda in a united quest for pure discovery. No process is perfect, of course, yet with every instance of deception—alleged or actual—the public's confidence in the honesty and integrity of those who regard themselves as researchers is shaken a little further.

In February 2010, the respected medical publication Lancet withdrew a controversial paper it had published back in 1998 on a possible relationship between the measles-mumps-rubella (MMR) vaccine, autism and inflammatory bowel disease. Over the years, ten of the study's twelve authors had distanced themselves from the article, removing their names from the document's conclusions as they acknowledged flaws in the research. Yet concerns raised by the paper's findings might have already impacted on the decisions made by numerous parents regarding their children's inoculations. Vaccination rates for the MMR vaccine in the UK subsequently dropped from 92 percent in 1996 to 84 percent in 2002, followed by an increase in cases of measles (56 in 1998, compared with 449 in 2006) and mumps (including an epidemic of 5,000 cases in 2005). While the media's reporting on this paper may not be entirely responsible for such outbreaks, it's hard to believe publicity of its findings didn't tip the scales as parents struggled with their choice. What's more, as the research was discredited in the popular media, vaccination rates climbed to 82 percent in 2004, up from 77 percent the previous year.

The study in question was based on observing a dozen children with Autism Spectrum Disorder, and while it did not explicitly conclude that there was a connection between their behavioral condition and the MMR vaccine, the lead author, Andrew Wakefield, nonetheless used the study to voice his concerns about the possibility of a connection and to recommend the combination vaccine be withheld until more research could be conducted.

Mistakes and deception can creep into even the most rigorous of studies, either through laziness, dishonesty or pure error. Images of assays are occasionally touched up somewhat generously, if not completely fabricated, while data is massaged, decimals dropped or results discarded entirely. Rarely is the intent to deceive, but rather to clarify or take a short cut. Often there is poor judgment in wording, or a neglect to report aspects of a method deemed to be trivial. Nonetheless, junk science of any kind presents a potential threat to the establishment of trust in reporting what is observed. By retracting a paper, Lancet was making a bold statement about its reliability as a journal.

In 2004, the British newspaper the *Sunday Times* revealed that a lawyer who was engaged in a court action against a manufacturer of MMR vaccines had recruited several of the children for the controversial study. The researchers were essentially being paid to find a connection between the vaccine and autism, prompting Wakefield to overstress the possibility of a connection without a suitable weight of research to back it up.

The consequences of this influence led to the UK's General Medical Council charging Wakefield with medical misconduct in early 2010, ruling that he acted "dishonestly and irresponsibly" in his research. That May he was struck off the UK's medical register.

The debate that emerged from the initial publishing of the article was understandably inflamed by passions on both sides of the argument. The public has long held a love–hate relationship with vaccinations. In 1853, the English government made it compulsory for the public to be vaccinated against smallpox, yet was later forced to produce a modified act in 1898 to take into account the conscientious objector. On one hand, the concept itself seems counter-intuitive, invasive and risky. Ever since the first drops of cow pus were scraped into lacerations in an attempt to arm the body against smallpox, people have been challenged to overcome their apprehensions and trust in the reasoning behind the action. Add to that fears of contaminants and "chemical" preservatives in modern vaccines, and it's easy to sympathize with such hesitation.

The British government's response to public concern over the MMR vaccine controversy included an attempt by the prime minister to present a sympathetic face, rather than impose a draconian law dictating a mandatory immunization program. Tony Blair stated in 2001 he believed the vaccine was safe enough for his son, presenting himself as a parent who also had to make the decision over their child's immunity. Yet his choice of words drew the ire of the media when he refused to categorically clarify whether his son had actually received the shots or not.

Where authorities face the responsibility of disease control, public mistrust can add fuel to the fire. Rio de Janeiro at the turn of the 20th century was a mix of glory and squalor. In 1902, persuaded by devastating epidemics of smallpox and yellow fever, Brazil's president instructed the city's mayor and director-general of public health to improve sanitation. Their response was a vast urban regeneration program, where entire neighborhoods were evicted and their dwellings demolished, replaced by lush gardens and building development. Mosquito exterminators and rat catchers were given authority to force their way onto premises in an effort to eradicate the vectors of many of the diseases.

By 1904, the director-general of public health was convinced public vaccination had to be mandatory, thereby passing a law entitling sanitation workers to enter homes and administer vaccinations, against the person's will if necessary. Regardless of the weight of the science behind the act, the public's trust in its government had vanished under the brutality of the law's implementation. Rumors were spread stating that women were required to undress for the procedure, fanning the flames of distrust even further. The result was six days of rioting as the population rebelled. Shops were looted, public transport was destroyed and even the cadets from the city's military college mutinied against the government.

The Brazilian vaccine revolt is an extreme example of public backlash to an authority's dictation. In hindsight it might be tempting to view it in light of its success in eradicating smallpox from the city. In democratically negotiating a course of action with the public, it's possible they might have opted against vaccination under the persuasion of fear and misinformation, leaving the government with the dilemma of imposing tyrannical health laws or risking ongoing epidemics. A third option would be to engage the public in an education campaign that aims to encourage cooperation. Yet education is a broad term that depends less on the quality of the information and more on the perception of the authority. "Vaccination works" can just as easily be "vaccination causes autism" and still be advertised under the guise of public awareness if the authority supports it.

The benefits of the philosopher's toolbox can be fully appreciated only in a democratic community, where its members share values in open discussion and skills in evaluative epistemology. Ideas cannot exist external to a cultural context, unbound by any one social framework or personal bias. Only when ideas persist in the minds of a diversity of tribes, repeatedly assessed against reality's constants and modified as new observations come to light with time, can they truly be declared useful.

THE TANGLED WEB

Long before there was the Internet, there was ARPAnet. And before there was ARPAnet, there was a diverse array of computer networks that couldn't communicate with one another. Initially, this wasn't considered as much of a problem. In 1962, American computer engineer Joseph Licklider was hired as director of the information processing techniques office (IPTO) at the US Defense Advanced Research Projects Agency. For him, it quickly became clear that distinct islands of computers would come to present more than just a mild inconvenience.

His department's resources included computer terminals connecting into MIT's Compatible Time-Sharing System, the System Development Corporation at Santa Monica and Berkeley's Project Genie, each of which required different sets of commands to engage with. Moving between them required physically walking to each terminal and entering individualized strings of instruction in order to pull out the necessary data. Licklider understood it would be far more efficient to have a single terminal with a single set of commands that was capable of digging into all three networks.

This desire for efficient connectivity began an endeavor to connect physically distinct computer networks under a single system. It wasn't until the end of the 1960s that IPTO director Robert Taylor would bring together a team to create the Advanced Research Projects Agency Network, or ARPAnet, which would connect the networks of four different research institutions across the United States.

While the concept of the modern Internet lies within ARPAnet's innovative communication strategies for sharing information by dividing it up into distinct packets, there persisted a problem of unifying the protocols under which different networks operated. Having a conference is more than just driving people to the same building and giving them name tags—they need to know when to shake hands, sit down and raise their arm when they have a question. The solution came in 1973 with the development of a common Internet protocol, or IP, which would be understood by the different networks. Once the computers in a network knew the so-called Transmission Control Program protocol, they could communicate with one another.

Today's Internet evolved from a practical need for distinct communities to share their information quickly and efficiently. In the 1980s, technicians at the European nuclear research organization CERN realized it wasn't enough to just be connected; there had to be a common way of presenting data. It was within this unified structure that the World Wide Web evolved, where information written in a specific computer code called hypertext could be interconnected within a branching structure.

Thanks to the proliferation of affordable, fast information technology, the web has become more than just a collaboration tool for researchers or a method of mailing data quickly over long distances. It has taken on characteristics of a distinct social sphere for many different tribes of people, where communication occurs within a digital territory that is commonly differentiated from "real life." With new tools for expressing and storing ideas, people were enabled to engage with one another across time and space using little more than a computer terminal and some technical know-how.

Importantly, the branching, globalised properties of the web defied the efforts of traditional authorities to exercise control over the flow of information. The web wasn't exclusive to a single class or nation. It could not be shut down by a law or have its plug pulled by an individual president's decree. The web had all the features of a perfectly democratic community, where anybody who could speak the language could contribute to the conversation without fearing censorship through an authority's filter. With rapid advances in software that simplified the need to know complex computing code, it meant virtually anybody could create a website or even write a blog with little education. At the end of 2010, the blog search engine Technorati.com was tracking just over 113 million blogs. According to their 2010 *State of the Blogosphere* report, half of the respondents used a free third-party blogging service. The anarchy of the web has produced a horizontal dissemination of information that could theoretically be accessed by anyone.

While the concept might have seemed utopian, the reality of several decades of web evolution has demonstrated that our social brains will gradually tend toward a hierarchical structure based on even the most meager differences in resource allocation. Even in a perfect democracy, there is a drift toward the control of how knowledge is communicated. Nations such as China have come to realize the potential of the web's reach, imposing filters screening what their citizens can access, with the governments of other countries such as Australia considering the instillation of similar control processes.

Without the physical impediments of a gatekeeper, information is still filtered by the practicalities of time. Given an overwhelming wealth of data, it becomes a numbers game as to what we read and what remains hidden from view. Posting a blog means nothing if nobody bothers to read it, while competition makes it even less likely that a large audience will stumble across the lone wallflower who is quietly whispering in the corner, regardless of what vital facts they have to offer. In the end, money still makes the difference between being somebody and being nobody. Economics is as important on the web as it is anywhere else in society. Thus it's the vying for attention and our need to interact in economical terms that means information has a greater chance of being spread through some channels than others. The web might be new, but the information market is no different today than when Gutenberg bound his first Bible.

According to Nielsen ratings, two-thirds of information searches on the Internet are performed using the search engine Google, with the remaining third spread primarily between four competitors. About 400 million people had a Facebook profile in 2010, with 35 million updating it daily. For all of its diversity, most people use the same selection of websites to buy books, interact with their friends, search, sell and gossip.

Wikipedia might be open to any single person to update and modify, yet editors are often driven to lock pages under threat of vandalism and administrators can censor edits. As within any community, regulations are born out of a need for balance between democracy and efficiency of interaction. For all of its diversity and freedom, there remain the same social forces within web culture as exist outside of it, controlling how information is spread.

THE DIGITAL TRIBE

"For this discovery of yours will create forgetfulness in the learners' souls, because they will not use their memories." If these words had been spoken today, the author could have easily been referring to the Internet. In an age when "wiki" and "Google" have become verbs and portable access to the great wealth of online libraries through the latest "i" device provides information regardless of time and place, we've become accustomed to the idea of searching for knowledge that we recall exists but don't remember in any great detail.

Given this portentous warning was at the hand of Plato on the subject of writing itself nearly 2,500 years ago, anxiety over the threat imposed by information technology is nothing new. Conrad Gessner—a Swiss naturalist—famously warned in his bibliography of 16th-century texts that an excess of information could prove not only confusing, but "harmful" to the population. If not for the fact that Gessner had written his landmark book *Bibliotheca Universalis* between 1545 and 1549, his reference to a potentially devastating flood of knowledge could easily have been a comment on modern blogs and tweets rather than the countless sheets of printed paper spat out of Gutenberg's press.

We've wondered how to deal with such an intimidating expanse of knowledge for the longest time. The digital age presents us with nothing fundamentally new. Each new medium we create merely presents us with yet another means of disseminating ideas to new groups of people, with the novelty being in the cost, speed, aesthetic and level of technological interaction. The invention of the written word made it possible for information to travel where spoken words couldn't, through both time and space. Printing made the written word affordable, sharing information with members of the community beyond the affluent gentry or the well-funded institutions like churches and universities. Wireless radio provided rural communities and illiterate folk with a means to access more information, while television added moving images to the broadcast of sound.

With each advance, however, communities have encountered new choices in where to find information. We evolved to cope with relatively small tribes of related individuals offering a few sources of information in the form of conversation, song, dance or oral tradition. The written word exposed us to tribes from the past or from distant lands. Printing opened the floodgates for information to spread far and wide, penned by authors who could now press out a manuscript without the necessary support of a rich patron. Today's digital age has created a diverse forum where any individual with access to the cheap, simple technology can be an author.

By the judgment of New York University psychologist Denis Pelli and graphic arts professor Charles Bigelow from the Rochester Institute of Technology, the number of authors in the world (somewhat arbitrarily defined as a person whose written word is read by at least 100 people) since the beginning of the 15th century has increased by a factor of ten every century. These days, a million people join the esteemed ranks of authorship each year, as the tenfold centennial increase has turned into an annual phenomenon thanks to our ability to reach more people through social networking tools.

Multitudes of people across all walks of life log on to the web to read their favorite blogs, check out a news feed or two, post their opinions on a forum or message board and throw a link onto their Twitter or Facebook network. Where once large volumes of people would drink from a narrow selection of information pools, the increase in the diversity of information sources gradually divides the audience. Traditional forms of media, from magazines to newspapers to commercial television stations, are being forced to deal with the reality of sharing a finite viewing population.

There is both danger and a boon in this rapid diversification of resources. On one hand, our information-hungry brains can satisfy their demands for answers at the touch of a button—regardless of the information's accuracy. Rather than conversing with a physician, for instance, people are turning to health websites to self-diagnose, often leading to further confusion and anxiety, not to mention occasional poor choices of medication or treatment. Without the skills to evaluate ideas critically we rely on the competition for our attention to subtly determine how useful information might be. Chain-letter e-mails containing lists of "Did you know?" warning of the end of the world in 2012 or informing the reader that Mars will be the size of the full moon this summer are often accepted as factual at face value.

Yet the accessibility of this information also makes it possible for a person to quickly evaluate how controversial a topic is, simply by identifying conflicting information or observing the discussions resulting from such disagreements. Lists of reader comments beneath a news article can serve to represent popular opinion, adding a new dimension to the previously one-sided affair of journalism.

In the so-called information age, the question of how to teach people to deal with multiple information authorities has become more important than ever.

"Better education" is the cry of many politicians, lobby groups and social movements who feel it is the answer to their pet problem of the modern world. But education needs to mean something beyond more resources or better paid teachers. Good thinking is not just a matter of possessing more knowledge or being "bright" or "intelligent," but rather of developing key learning skills in order to evaluate information efficiently, identifying the useful from the useless. Conversely, bad thinking is not a reflection of stupidity or insanity, of broken neurology or willful ignorance.

Most importantly, education is not simply a proposition of facts. In a world of competing voices, it's not enough to be just another authority and think it will serve the young citizen well in their future years.

Critical literacy—the ability to evaluate the context of what you read or hear to gauge its accuracy—will rely on different sets of evaluative tools than in the past. Rather than be repositories of knowledge in the form of graffiti-marked textbooks or decades-old BBC nature documentaries, schools will need to promote methods for students to navigate their way through the confusing digital forest that surrounds them, identifying efficient ways of accessing the most accurate information with search tools while recognizing how to make use of an overwhelming ocean of data.

The question of how to engage citizens in science must move beyond a focus on the dictation of facts, figures, beliefs and theories, and have as a greater priority the ability to engage with others in the community to evaluate the knowledge they encounter.

Within us all, our tribal brains tick away, making use of social networking tools in an effort to share and evaluate new information. There will always be hierarchies controlling the flow of information due to our social machinery. The way we use these tools will determine how we cope in a noisy future where a cacophony of information surrounds us.

Chapter 8

THE PROGRESSIVE HUMAN

What will intelligence mean in the future?

In solving a problem of this sort, the grand thing is to be able
to reason backward. That is a very useful accomplishment,
and a very easy one, but people do not practise it much. In the
everyday affairs of life it is more useful to reason forward, and so
the other comes to be neglected. There are fifty who can reason
synthetically for one who can reason analytically.
— *A Study in Scarlet*, Arthur Conan Doyle, 1887

THE MAGICIAN AND THE KNIGHT

Although technically uttered by a fictitious detective known for
his extraordinary ability to think logically, it was the British
author Sir Arthur Conan Doyle who dreamed up these insightful words
on behalf of his famous crime-fighting creation, Sherlock Holmes.

Undoubtedly, Doyle was a man who possessed a healthy respect for
the mind. On graduating from studying medicine at the University of
Edinburgh, he became a ship's doctor before joining a classmate in setting up a medical practice in Plymouth. Their troublesome partnership
was short-lived, leading Doyle to open his own clinic in the English seaside town of Southsea, where due to a shortage of patients he penned
fiction to while away the hours. It was there in 1887 that he not only
threaded together the narrative that would be his first Sherlock Holmes

book, *A Study in Scarlet*, he began to express a passing interest in paranormal phenomena.

It wasn't until the end of the First World War that his curiosity for the supernatural evolved into full religious faith. He'd abandoned his Catholic roots when he was still a young man, confessing to agnosticism, yet it was into the burgeoning Spiritualist movement that he was reborn.

Spiritualism developed within the superstitious communities of rural New York during the 1840s and quickly spread across the country and over the Atlantic into Britain and Continental Europe. Inspired by descriptions of the spirit world by the Swedish scientist Emanuel Swedenborg and the teachings of German celebrity (and self-proclaimed healer and spirit communicator) Franz Friedrich Anton Mesmer, a loose denomination formed around the shared belief in the divine power of clairvoyance. Ironically, it may have been Doyle's rationality and regard for evidence that led him to embrace this religion over all others.

The first decades of the 20th century were full of tragedy for the author; 1906 saw the death of his first wife, Louisa, from a bout of tuberculosis, while his son Kingsley and brother Innes were to succumb to pneumonia in 1918 and 1919 respectively. Two of his brothers-in-law and two nephews also passed away in the years following the First World War. Where most church communities rely purely on faith in an eternal life after this one, Spiritualism tempted Doyle with tangible proof of a supernatural existence he could experience first hand. His values in science might have been heavily compromised by a deep desire for there to be something beyond the natural world; however, it was this contrast that would lead him to form a turbulent and yet cherished friendship with a man who would become known as the world's greatest escapologist.

The American magician Ehrich Weiss—or, as he was more famously known, Harry Houdini—developed a more personal interest in the paranormal following the death of his mother, Cecilia, in 1913. Already well acquainted with theatrical stage performances involving mediums and clairvoyants, Houdini even had the good fortune to meet with one half of the Davenport Brothers and learn a trade secret about the use of their bindings in their popular "Spirit Cabinet" routine. The act involved the brothers being tied securely in their chairs within a large wardrobe, which on closing would erupt with a chaos of knocking and music from horns and drums. In the name of publicity, the brothers were never forthcoming about the true nature of their act, choosing to let the audience's imaginations run their course as to whether it was mere trickery or something less mundane and more mysterious.

Yet a single fraudulent act wasn't enough for Houdini to presume all mediums were fakes. He remained hopeful that there was something to the phenomenon and never ceased in his search for a genuine psychic.

On touring the British Isles in 1920, he sent Doyle a copy of his book *The Unmasking of Robert-Houdin*, in which there was a reference to his meeting with the medium/showman Ira Davenport. This innocent act was the first in an ongoing correspondence between the two men where Doyle would gruffly dismiss any possibility that the Davenports were but mere conjurers, leaving Houdini to deftly veil his skepticism in order to not cause offense while continuing to draw out more of the author's opinions.

Finally the pair met in person at Doyle's Sussex country residence, where Houdini joined the writer and his second wife and discussed shared interest in the paranormal over lunch. Doyle ended the afternoon promising to introduce Houdini to a number of mediums who were most undoubtedly as genuine as they came.

Needless to say, the magician found their acts to be less than convincing. An organization called the Society for Psychical Research hosted a series of séances that Houdini was invited to observe. They included several mediated by a French woman named Eva C, renowned for her talent for excreting blobs of pale, gelatinous goo referred to by Spiritualists as "ectoplasm" from various parts of her body. The researchers of the day were wise to some of the more amateur tricks conducted by fraudulent mediums, so would sew them into tights to prevent them from using their lower orifices as hiding places for any such material and feed them cake or bread to make it inconvenient for them to hide anything in their stomach. Of course it still left a few less obvious possibilities for a clever trickster to discreetly tuck away slivers of bleached cheesecloth, wads of chewed paper or whatever else might have served to represent paranormal secretions.

Nothing of note happened during Eva C's first two performances.

Throughout the third night, small amounts of whitish goo were produced, which the observers carefully noted. Houdini made no guesses as to the nature of the materials, given they disappeared before he could ask for a sample. Nor did he attempt to determine where she might have hidden them (let alone to where they might have vanished). However, given his understanding of sleight of hand, he believed there was no need to resort to the supernatural to explain what he saw.

To the magician's amusement, he occasionally encountered personal accusations of possessing supernatural talents. The president of the British College of Psychical Science, JH McKenzie, had written of Houdini, "This ability to unbolt locked doors is undoubtedly due to Houdini's mediumistic powers and not to any normal operation of the lock. The effort necessary to shoot a bolt from within a lock is drawn from Houdini the medium, but it must not be thought that this is the only means by which he can escape from his prison."

As the seasons slipped by, friendly letters continued to be exchanged between Houdini and Doyle, neither managing to convince the other of their position but taking delight in the sharing of views. Houdini rarely made an effort to be overly critical toward Doyle's claims; even when he received a letter describing with childlike glee a collection of photographs taken by two young sisters depicting dancing fairies, he resisted scoffing or ridiculing his friend. Instead, he would politely state his own reservations on any topic to do with the paranormal, stating stoically that he had yet to be convinced.

There was only one occasion where Houdini made a serious attempt to demonstrate to his visiting friend how easy it was for anybody to be fooled. On a visit to Houdini's home with Houdini's friend Bernard Ernst, Doyle was invited to participate in a simple demonstration. The three men adjourned to a sitting room where a free-standing slate was propped up on an easel and a plate holding three cork balls sat on a table. As later recounted by Ernst, Doyle was free to examine the room's contents until he was satisfied all was as it appeared. He was then asked to go some place where he felt assured of privacy, at which point he was to write on a piece of paper any words that came to mind. The writer did so, traveler several blocks down the street before pausing to write the Biblical phrase *"Mene, mene, tekel upharsin"* (Daniel 5:1—The words written on the wall at King Belshazzar's feast, translating roughly into the coin denominations *"mina," "shekel"* and *"peres,"* or half-mina). On his return, Doyle was asked to dip one of the cork balls in white ink, where Houdini then took it and placed it against the slate. To the author's amazement, the ball proceeded to roll about, spelling out the very words he had written.

As any magician would know, there was far more to the story than this simple narrative. Subtle misdirections and missed details hide the story's true events. However, if Houdini had hoped for Doyle to simply take his word on the stunt being little more than a clever deception, he was sadly mistaken. The author was now convinced beyond doubt that his magician friend was secretly assisted by supernatural powers. Houdini could do nothing short of revealing the trick's secret, which, reluctantly, he wasn't prepared to do.

On a weekend in June 1922, Doyle would engage in his own attempt to persuade the magician to believe in the paranormal. As he and his family toured the United States, they invited Houdini's family to visit them in Atlantic City. In the evening, Doyle's wife, Jean Elizabeth, surprised Houdini with an impromptu séance in an attempt to contact the magician's deceased mother, Cecilia. The trio sat around a small table, Doyle handing Houdini the notes his wife scribbled as she moaned and writhed dramatically. Fifteen pages of notes were scrawled over the next few hours, communicating a mother's love and adoration but no significant personal details that might have caught Houdini's attention.

Although he admitted nothing would have pleased him more than to once again speak with his dear mother, the evidence was far from compelling. Cecilia's birthday had been but the day before and there was no mention of it in the séance. Peculiarly, the first mark Jean Elizabeth made was a cross; Cecilia was Jewish. But most damning of all, in spite of five decades living in America, Houdini's mother could not speak nor write a single word of English.

By October, the pair's relationship began to slowly dissolve as it was dragged through the mud of a public debate on Spiritualism. Houdini's involvement in exposing fraudulent psychics—once respected by Doyle for cleaning out the offending "fakes"—was becoming a sensitive issue as his focus turned on those the author felt were genuine. Both men increasingly found themselves misquoted by the media, leading to increasing bitterness and anger.

Sir Arthur Conan Doyle would outlive his old friend by four years and maintain throughout them that in spite of the magician's skepticism, Houdini's extraordinary talents could be nothing but superhuman. In an essay published in the Strand magazine, Doyle wrote, "Who was the greatest medium-baiter of modern times? Undoubtedly Houdini. Who was the greatest medium of modern times? There are some who would be inclined to give the same answer."

The two men were the most unlikely of friends. Doyle was a well-educated British physician who found fame through his talent for the written word; Houdini emigrated from Hungary to America with his family when he was four and had little in the way of formal academic tuition. Both were considerably intelligent, worldly men who longed for evidence of a world beyond the physical. But their smarts differed in fundamental and significant ways, leading one to conclude that fairies existed in the bottom of the garden and that spirits spoke through the living, and the other to see nothing supernatural no matter how hard he looked.

THE FALSE DICHOTOMY OF BELIEF

According to a 2008 social study conducted by researchers Martin Bridgstock and Kylie Sturgess, more than half of the Queensland population are estimated to believe that through mental effort the human mind can heal the body. A little over a third believe that the souls of the dead can come back as ghosts, and almost 30 percent think that the stars can influence how a newborn's personality will form. By the same token, over 70 percent of Queenslanders aren't convinced that it's possible to mentally communicate with the dead, and the same percentage give little or no credence to the belief that aliens have visited Earth at some time in the past.

Regardless of which group you sympathize with most, you have to agree that both sides can't be equally correct. This gives rise to the question: what leads a person to choose which side of the fence to stand on when confronted with the question "what do you believe?"

Both Houdini and Doyle were willing to accept there was an afterlife. Both men were relatively well informed, articulate and logical. Yet there was a rather simple and yet fundamental difference in how each of them approached new information they personally encountered.

Houdini was a critical thinker. Every idea presented to him was tested with a strong expectation that it could very easily be wrong. His background as a magician made it clear that things weren't always as they might appear to be, and a single peek behind the curtain was all it took to expose an illusion. Rather than look for reasons to continue to believe, he approached all ideas with an understanding that the harder it was to identify a fault the more likely it was going to be a useful way of describing the world.

Doyle, on the other hand, sought confirmation for what he already had his heart set on believing. Rather than consider why each new piece of information might be misleading, he found it difficult to comprehend how a man of his intellect could be deceived so easily. Doyle was undoubtedly a knowledgeable individual, yet bad ideas can easily be supported with careful selection of even the most solid facts and figures.

We're not all pure Houdinis or absolute Doyles, even though we might identify closely with aspects of either. In spite of surveys that appear to state otherwise, belief versus non-belief in the paranormal, pseudoscience or the supernatural is a false dichotomy. The human population cannot be sharply divided into opposing categories for all ideas. There is nothing to stop you from claiming fairies exist while scoffing at the likelihood of ghosts. You might think aliens have visited Earth in the distant past but not recently, or believe in the power of *qi* but not the efficacy of acupuncture. It's possible to think there is an all-powerful deity but not an afterlife. "Belief" itself is far from a discrete line we cross. A person might not fully embrace the possibility of clairvoyance but may have a niggling feeling that "something" mysterious might be going on when they have a sense of déjà vu. Astrology might be regarded as nonsense, but a person might still entertain the notion that it's possible to categorize personalities according to a similar system using blood type or eye color, or think that a person's behavior can still somehow be influenced by the positions of planets.

On the flip side of the coin, self-declared skeptics can arrive at their conclusions independently of the application of any critical thinking whatsoever. Just as a person brought up in a religious household might find it difficult to believe their family is mistaken in their belief in a god, one brought up in an atheist household can reject the possibility of a supernatural creator out of habit, or a subconscious respect for the beliefs of their parents and siblings.

Surrounded by doubters, people can easily be influenced into swallowing the conclusions of their tribes without having arrived at them through independent thought.

Yet just because there is a rainbow doesn't mean individual colors don't exist. We all contain shades and hues of Houdini and Doyle in certain ways.

A wide variety of studies over the years have attempted to identify the reasons why some people lean toward one idea over another. Many of the studies succeed in identifying correlations, suggesting the formation of a particular belief might be the result of a complicated spectrum of interacting factors rather than any one cause.

Going by a 2001 survey by the US National Science Foundation, more women than men believed in extrasensory perception (ESP), while the balance was roughly even when it came to alien visitation. Possessing a college-level education seems to dramatically decrease an acceptance of astrology while having surprisingly little impact on a belief in other phenomena. According to the same survey, eight out of the thirteen suggested paranormal topics attracted more believers during the 1990s. Only demonic possession experienced a decrease— the remaining four fields remaining more or less the same.

Age, gender, cultural background, socioeconomic status and level of education all seem to have some impact on the popularity of specific paranormal or pseudoscientific beliefs. Runs of certain movie themes or television programs—be it poltergeists, mediums or alien visitation—affect the perception of some paranormal beliefs. In the past, mass hysterias could be provoked by selective reporting in newspapers, as with sensationalized articles describing masked intruders or muggers armed with poisons.

By far the biggest influences on our belief in the paranormal, however, are those personal events that seem to defy a straightforward explanation. Close to half of people who believe in ESP cite their own confusing experiences as the reason why. Whether it's a sense of premonition, a dream that appears to predict future events or the vague feeling that we know something about a person we've never met before, these sensations often provoke some people into embracing the paranormal as a likely explanation.

We're inspired from all angles to believe certain things, whether it is through friends within our local neighborhood, students or teachers within our schools or the media we read, hear and watch. All of these influences work with our own personal experiences and values to produce a degree of confidence in whether something is likely to be true or not.

Ideally, we would all be perfectly rational Houdinis, accurately picking truth from nonsense. In some ways, most of us are under the impression that we do well at this already. Psychologists Justin Kruger and David Dunning from Cornell University demonstrated in the 1990s what many within their academic field had long suspected: most of us describe our cognitive abilities as above average. When scored on tests involving humour, logic and grammar, a significant number of those who scored the lowest overestimated how well they had done. Conversely, those who did the best in the study tended to underestimate their performance. In the words of the philosopher Bertrand Russell, "The fundamental cause of the trouble is that in the modern world the stupid are cocksure while the intelligent are full of doubt."

Paradoxically, having your belief shaken might even cause you to cling to it even tighter. A 2010 study conducted at Northwestern University in Illinois by marketing researchers David Gal and Derek Rucker tested an earlier study by the psychologist Leon Festinger that those with supernatural faith tend to proselytize more vigorously when their faith is challenged. Festinger had also demonstrated that people tend to change their beliefs to suit their behavior, rather than change behaviors to suit a change in belief. Gal and Rucker's supporting study looked at an interesting phenomenon where views expressed in writing using our least favored hand lower our confidence in what we've written. In this case, students who penned an opinion with their weaker hand weren't as confident in their response to a question on animal testing as those who were permitted to use their favored hand. Oddly, this contrasted with their outward behavior; when asked to write a persuasive writing piece on the topic, their responses were longer. Doubt, it seems, only serves to drive us to work harder to reinforce what we have already accepted as true, decreasing the chance we'll fracture the beliefs shared within a tribe.

Faced with the aggressive demands for reasons by our information-hungry left hemisphere, it's easy to see why a sizeable percentage of the population are unable to admit they honestly don't have an explanation without invoking the supernatural, which makes them put so much effort into arguing their case. What, then, allows others to resist? Are our brains fundamentally so different from one another's, or have we simply learned to use its tools slightly differently?

At some point in our lives, we are forced to deal with the limitations of human experience. In spite of our brain's innate tendency to favor emotion over reason, some tribal settings are more conducive to openly expressing critical-thinking skills while suppressing the innate urge to repeat the beliefs of peers. Whether it's a family environment, role model, school or a mix of numerous community interactions, some individuals will sympathize more with scientific values and learn to develop the thinking tools appropriate for evaluating information.

Interestingly, merely knowing the products of science isn't enough to develop such epistemological values. There is no relationship between having done science as a subject at school and having good critical reasoning skills, for instance. However, when education involves the role modeling of effective thinking skills in addition to merely learning science facts, students invariably develop the values and abilities required to critically address new ideas. Learning how our tribal thinking facilitates the creation of knowledge is increasingly necessary in a scientific world.

Our understanding of the universe is becoming increasingly complicated, as is the technology that emerges from it. Identifying practical solutions within a realm of possibilities is becoming more difficult as time passes. There is at once a glorification of the future and a fear of a potential dystopia that technology will bring. Stories of exponential progress and technological "singularities" emerge out of speculations of artificial intelligence providing super-intelligent computers capable of solving our most pressing problems, if they don't eradicate humanity in the process. As science becomes more complicated and less intuitive, technology will become the modern miracle we pray for in a time of need.

THE GLOWING WIND

With a name like a *Star Wars* droid, ADE 651 sounds like it looks—a complicated piece of technological wizardry. One half of the device is held in the hand like a pistol, with a metal rod jutting forth where you'd expect a comical flag labeled "Bang!" to unroll. It's connected by a lead to a small black box, in turn strapped to a belt wrapped around the user's waist. According to its UK distributor, Advanced Tactical Security & Communications, ADE 651 "incorporates electrostatic ion attraction technology to target the specific substances," namely materials such as explosives, drugs or human remains. Interpretation? It allegedly sniffs out what you want to find by identifying the particular charge and shape of any molecules that item happens to shed into the air.

Whether the mechanism is pure science fiction or relies on a plausible concept is beside the point—according to tests performed at the US National Explosive Engineering Sciences Security Center, the device itself works no better than random chance. It's about as good as taking a wild stab in the dark.

Not to say that minor issue stopped the Iraqi government. According to the New York Times journalist Rod Nordland, the government has spent approximately $85 million since 2007 on a supply of units for their roadside checkpoints. The country's head of the Ministry of the Interior's General Directorate for Combating Explosives, Major General Jehad al-Jabiri, dismisses any possibility that ADE 651 is nothing more than an expensive paperweight by stating he knows more about explosive devices than anybody else in the world, and their purchase has proven effective at reducing roadside bombings.

We're increasingly forced to deal with such magical "black boxes" in our everyday lives. While most of us can readily recognize a microchip, few people could explain how an assortment of electrical components works to interpret strings of code that result in a working computer game. It takes little more than a high-school understanding of optics and chemistry to grasp how an old-fashioned movie projector and its film produce a moving picture, yet how many high-school graduates could today explain the full process that makes a plasma television work?

As technology progresses and our discoveries regarding the universe expand in number and detail, the average person is less able to rely solely on their personal knowledge to determine whether something is feasible or not. No matter how magical something seems, most of us are faced with it being possible through some fantastic yet plausible twist of technology. The weirdness of quantum mechanics, with suggestions of particles appearing and vanishing, teleporting and communicating instantaneously across time and space, and existing as both waves and particles at the same time, serves to further confuse matters.

The 17th-century British physicist Isaac Newton felt light was best described as streams of tiny particles. This explained reflections well enough, but the problem was the bending of light as it passed through different materials, a phenomenon referred to as "refraction." Why would a stream of particles change direction as it entered and exited a prism of quartz? Newton suggested it had something to do with an increase in gravity making the particles accelerate as they hit dense, transparent substances, thus diverting their paths.

Of course, if light truly was a wave, there would be no problem explaining either refraction or diffraction. Waves change direction as the medium changes its density, after all. Yet a wave is a pattern of movement in a medium, such as water or air. If an ocean of intangible fluid permeated the universe, waves of light from distant stars might carry across the vast distances of otherwise empty space. There would be no need to use particles as an explanation at all, in which case. The hypothetical fluid was called "luminiferous ether," and it remained a viable theory for explaining light for several centuries.

Throw a rock into a pond and the ripples will radiate away from where it splashes into the water. Take a cue ball in a game of pool and roll it along the felt until it bounces from one of the side rails. The argument on which analogy best describes light goes as far back as the 17th century, with a debate over whether light behaved like bullets from a machine gun or waves through a water tank. The eventual answer counts as one of the greatest twists the history of science has ever seen.

A Dutch physicist by the name of Christian Huygens observed the manner in which light bounces from a mirror, spreads through a lens and makes shadows on the wall and figured it must have been a wave passing through the air just like sound.

Later, Newton published his belief that since a beam of light ricochets from a mirror along the same geometry as a ball bouncing off a wall, and travels in a straight line in one direction, light must be made of particles. The scientific community liked Newton's so-called corpuscular theory of light, and for the next century there was little argument.

An experiment conducted in 1803 by another English scientist, Thomas Young, demonstrated that as light passes through a slit it spreads out, just like a wave and not at all like a stream of particles. With that, the old argument was reignited. A second slit paired next to the first caused the light beams to interfere with one another, making some spots brighter and some spots darker. If light truly was a stream of particles, two similar-looking patches of light should be seen.

Young's experiment turned the physics world on its head, and after the French scientist Augustin Fresnel expanded on his conclusions, the world seemed satisfied to accept the Huygens-Fresnel wave theory of light.

A century after Young's double-slit experiment, a German named Max Planck stumbled onto a problem with how an object emitted radiation. Going by his sums, the only way he could solve the problem was to assume energy was actually segregated into neat packets, or "quanta." Einstein clarified this by dividing light into packets called "photons." In other words, light was best explained as a particle once again.

So which is it? Suffice to say, light behaves neither like a cue ball nor a ripple on a pond, but has properties of both. There just isn't a simple analogy we can use. On the tiniest of scales, it seems that our personal experience of the universe has no equivalent.

In order to understand the world, our brain makes use of analogies. It constructs knowledge from the pieces of existing ideas, slotting beliefs together into a scaffold and incorporating novel observations, just as we use metaphors to explain inchoate concepts. Yet in circumstances we have little experience of—like the quantum universe—we struggle to find good analogies. Our brains are best suited to a classical universe of waves and particles and not its unfamiliar fabric where subatomic particles pop in and out of existence and form connections that ignore time and distance.

As the 19th century drew to a close, the argument was in full heat and the search was on for evidence. In 1887, a duo of American physicists by the names of Albert Michelson and Edward Morley came up with a clever way to test for the presence of luminiferous ether. Presuming it would behave like a fluid that filled the universe, ether should blow on the Earth as the solar system sailed through the galaxy. What's more, they realized the strength of this breeze should change as the Earth rotated and orbited the sun, sometimes moving with the current and sometimes against it. If they could detect such variations in the speed of light, it would help support the idea of the light forming as ripples within a vast fluid drifting through space.

Timing a pulse of light isn't as simple as standing at one end of a room with a stopwatch and a flashlight. During the 1670s, a Danish astronomer by the name of Ole Rømer had provided data on variations in the orbit of Jupiter's moon Io, determined to have been caused by differences in the time its light took to arrive on Earth. From this the Dutch physicist Christian Huygens calculated that light traveled 16.6 Earth diameters per second, or only about 665,000 kilometers per second. Unfortunately, he misinterpreted Rømer's notes and came up with an answer that was too big. Other astronomers made far better attempts—in 1809, Jean Baptiste Joseph Delambre determined it took 8 minutes and 13 seconds for light to zip from the sun to the Earth. This distance varies somewhat during the course of a year, but it gave a good estimate of the speed at roughly 300,000 kilometers per second.

Even with the best tools of the 19th century, however, scientists could only measure light to within 5 percent accuracy—nowhere near efficient enough to find the subtle fluctuations expected if light was indeed carried by ether.

Michelson's solution was a stroke of genius. He would use the fact that waves, including light, create unique patterns when they interfere with one another. Using mirrors he split a ray of light into two beams, allowing one to line up with the suspected ether current, and the other beam to cut across its path. Both were then reflected to match up again on a detector. If the beam cutting across the supposed ether was slightly slower, the rays would fail to line up, creating a distinct interference pattern. Any disparity would show that a medium did, in fact, exist and that light was, in fact, a wave.

Tiny differences in the timing due to minor differences in the mirrors and the shape of the glass were inevitable, but with a few tweaks it was deemed good enough to detect significant variations caused by any mysterious wave-carrying medium.

Their results serve as a reminder of how experiments that seem to fail are no less important than those that validate our expectations. Although there was a small interference pattern, it was approximately one-sixth of the size they'd predicted. If ether was out there, it was indistinguishable from the errors expected from their method.

Unconvinced, both men went on adjusting the experiment for many years, setting up special tents so the universe's gentle wind wouldn't be impeded by any solid walls or adjusting their apparatus to reduce unwanted errors. Others joined in, and for decades physicists attempted to reduce the margin of error down low enough in the hope that the subtle whispering of ether against the Earth could be felt by their quivering beams of light.

There remains virtually no doubt today that light has properties of both particles and waves. Yet we are no closer to knowing precisely what a light wave is than we were in the 19th century. More strange still, its speed seems to have no relationship with its surroundings. Shooting a laser from the front of a speeding train will not make the light flow faster than about 300,000 kilometers per second, no matter where you stand. The best we can manage is to scatter a forest of matter in its path to interrupt its journey and slow it down as each atom absorbs and re-emits it. Light's substance has no brakes and no accelerator. In fact, time itself will give way, warping and slowing, before the speed of light would dare vary by a fraction of a millimeter per second.

But what gives rise to the wave-like properties of a beam of radiation? Without a fluid to ripple through, it is a rather perplexing mystery. A substance like luminiferous ether would still be a nice idea to use to answer this question, neatly explaining a troublesome problem. Without evidence of its existence, however, we can make no assumptions of how such a medium would function. We know nothing of its flow, its density, or what it might be made of. In fact, it simply might as well not exist at all, for all the good such a hypothesis would do.

Interestingly, there is reason to believe that space isn't quite as empty as we presume. Far from being a complete absence of anything at all, we can imagine the very canvas that the universe is painted on as a turbulent medium that interacts directly with the particles and forces we call matter. However, its nature is vastly different from the gases and fluids we observe in our everyday lives. Space-time does not behave like water or air. In the future, we might make some significant comparisons between it and the defunct luminiferous ether hypothesis; however, for all purposes the notion of a universal fluid through which light can ripple fails in its attempt to describe the complexities of a beam of light.

In exploring the quantum universe we've encountered concepts that are beyond our current ability to comprehend. Try as we might, we can't conceptualize wave-form collapses of fundamental particles in the same way we can picture a bouncing ball. We still don't know how gravity relates to forces of electromagnetism and the nuclear forces. The universe is at once confusing, majestic, beautiful, logical and incomprehensible. And yet something in our tribal wiring makes it impossible for us to stop trying to understand it.

BRAINS, BELIEFS AND GOOD IDEAS

Science is a most fortunate accident. The human brain evolved features that enabled individuals to cooperate with one another in a harsh and unpredictable environment; yet those same features of language and logic that allow us to negotiate the politics of human relationships also allowed our ancestors to imagine the surrounding world in terms of ideas rather than just stimuli. Evolution never favored the scientific as much as it did the social, finding the key to survival in small tribal groups living a nomadic lifestyle.

Today's world is vastly different from the one we evolved in. Civilization has progressed faster than the clumsy, trial-and-error process of natural selection, creating a social landscape foreign to the intimate one our brains adapted to thrive in. Useful information has become a valuable resource allowing communities to provide their people with the means to better cope with the challenges of a changing world. Science has become a way of identifying that information. However, this process isn't the result of any single trick in the brain of some ancient savannah ape, but one that emerges from the complex social dynamics of colliding cultures. No single person can be a scientist in isolation. Science cannot exist in a single brain.

South of the Thames in the inner London borough of Southwark a patch of the 19th-century medical history remained hidden behind a wall of bricks for nearly 100 years. Closed in 1862 to make way for the Charing Cross Railway, Saint Thomas's Hospital and its medical contents were merely walled up and abandoned for a new site across the river. On being rediscovered in 1957, it provided a glimpse of a world that had long been left behind, of a time when surgery was quick and dirty and pus, sepsis and fever were par for the course.

A small, wooden amphitheater with rickety banisters forms a horseshoe around a narrow table in what was once, long ago, a storage space above a church and a refuge for the sick and poor. To one side there is a doorway that opens to a room beneath exposed rafters, desiccated stalks and leaves hanging from nails alluding to its original use as an herb garret. The theater's tiered observation stands serve as a reminder of how the phrase "operating theater" came about. Directly over the table there is an iron candelabra supplementing whatever weak light might filter down through the skylight, while overlooking the scene is a plain sign stating in Latin *"Miseratione non mercede"*—"For compassion, not pay." Visitors are told that the floorboards were found packed with sawdust to soak up the inevitable torrent of blood that would flow during the butchery that served as a surgical procedure.

How many limbs were sawn off before an audience in this hall? How many kidney stones did physicians cut out of their red-faced, screaming patients? What manner of fistulas and hernias did they slice and stitch? How many lost their lives under the surgeon's blades? The sanitized air of the museum is no longer marred by the scent of rot and old body fluids. The doctors in their leather aprons stiffened by the residue of their work would be unrecognizable today, bearing a closer resemblance to a worker at an abattoir than the gowned and masked surgeon we've become accustomed to seeing hovering over a quiet, sleeping patient.

What is amazing to consider is not the weight of death tainting the air in that room, but the fact that people even survived at all. It is a testament to the strength of human biology and its ability to persist through all but the most horrendous of circumstances. Yet given the perspective of over a century of discoveries that include drugs such as anesthesia and antibiotics, it's easy to take for granted our chances of making it to a ripe old age while forgetting that it has not always been this way.

In the moments we do reflect on the history of medical progress, we can't help but define it as a series of "Eureka!" moments, a lineage of philosophers who had their own metaphorical apples fall on their heads in instances of enlightenment. Modern storytelling presented by a sensation-loving media presents us with a simplified view of science as a punctuated equilibrium of events—dramatic occasions where the world pivoted toward a bright new future in a flash of inspiration. A chance examination of a rogue fungus on a Petri dish is singled out to define the moment penicillin made its scientific entrance—forgotten is the rigorous testing that followed, not to mention the heated debates, the skeptics and the fear-mongering. Given hindsight, science is a path to an absolute conclusion, where the endpoint is discovery.

In a letter to fellow philosopher Robert Hooke, Newton famously wrote, "If I have seen further it is only by standing on the shoulders of giants." Science is not a pyramid of brilliant minds, but a tree where ideas flourish or die as they are passed from one branch to another.

The Darwinian struggle between good and bad ideas, like evolution itself, is far from perfect when viewed in a greater historical context. In contrast to how it often appears, the process is slow and often fails to meet our brain's demand for absolute answers. Pseudoscience is an inevitable stain that will weather any attempt to be washed away, by virtue of our own humanity and its origins in our social brains. Yet we only need look around us to see that science does, eventually, progress in the face of all odds.

Homo sapiens never evolved with the express purpose of science in mind. Like all animals, our biology serves only one purpose—survival in a changing environment. It's not without a hint of irony, however, that the sum of talents that allowed us to strengthen our personal bonds through stories of gods and monsters are the very same talents that allowed us to reason them out of existence, permitting us to slowly hear the story of the universe itself.

ACKNOWLEDGMENTS

There have been a number of people who have contributed to the production of this book, either directly or through their support. First and foremost is Jasmine Leong, whose honesty, keen editorial eye, friendship and faith in my abilities as a writer have been constant sources of motivation. Tom Dullemond, Kylie Sturgess, Martin Bridgstock and David Shaw have been vital sources of critical feedback on all manner of subjects, while Barbara Drescher deserves special thanks for her constructive advice and expertise in cognitive psychology. And much love to my partner Liz, the most tolerant wife known to humanity, whose support and sandwiches have sustained me as I've buried myself in writing and research for months on end.

Tribal Science commenced life as a lecture in 2008 for the Brisscience program, titled "A Wolf in Sheep's Labcoat." If not for Joel Gilmore inviting me to talk on the subject of pseudoscience, it's unlikely that I would have dug so deeply into the relationship between social thinking and the human practice of science in the first place.

Lastly, my deepest appreciation goes to my eternally patient agent, Pippa Masson, publisher Alexandra Payne and the talent of editor Rebecca Roberts, whose dedication and clear vision have been invaluable.

REFERENCES

CHAPTER 1. THE STORYTELLING MONKEY

Chapman, M, *Constructive Evolution: Origins and development of Piaget's thoughts*, Cambridge University Press, Melbourne, 1988.

Cunnane, S, *Survival of the Fattest: The key to human brain evolution*, World Scientific Publishing Co., Sydney, 2005.

De Saussure, F, *Course in General Linguistics*, Open Court Publishing Co., Chicago, 1972.

Dunbar, R, Grooming, *Gossip and the Evolution of Language*, Harvard University Press, Cambridge, MA, 1998.

———. "The social brain hypothesis," *Evolutionary Anthropology*, 6:5, *1998*, pp. 178–89.

Finlay, BL & RB Darlington, "Linked regularities in the development and evolution of mammalian brains," *Science*, 268, 1995, pp. 1678–84.

Gazzaniga, MS, *Human: The science of what makes us unique*, HarperCollins, New York, 2008.

Gould, SJ, *The Mismeasure of Man*, WW Norton & Company, New York, 1981.

Hood, B, *Supersense*, HarperOne, New York, 2009.

Kenneally, C, *The First Word: The search for the origins of language*, Viking, New York, 2007.

Macmillan, M, *An Odd Kind of Fame: Stories of Phineas Gage*, Massachusetts Institute of Technology, Cambridge, MA, 2000.

Paterniti, M, *Driving Mr Albert: A trip across America with Einstein's brain*, Dial Press, New York, 2001.

Rózsa, L, "The rise of non-adaptive intelligence in humans under pathogen pressure," *Medical Hypotheses*, 70, 2008, pp. 685–90.

Taylor, M, BS Cartwright & SM Carlson, "A developmental investigation of children's imaginary companions," *Developmental Psychology*, 29:2, 1993, pp. 276–85.

Wolford, G, MB Miller & M Gazzaniga, "The left hemisphere's role in hypothesis formation," *The Journal of Neuroscience*, 20:RC64, 2000, pp. 1–4.

Woolley, AW, CF Chabris, A Pentland, N Hashmi & TW Malone, "Evidence for a collective intelligence factor in the performance of human groups," *Science*, 330: 6004, 2010, pp. 686–8, published online, DOI: 10.1126/science.1193147, <www.sciencemag.org/content/330/6004/686.full>, accessed November 17, 2010.

CHAPTER 2. THE CREATIVE SERPENT

Allen, RE, *Greek Philosophy: Thales to Aristotle*, 3rd edn, Free Press, New York, 1991.

Diogenes Laërtius, *Lives of the Eminent Philosophers*, trans. Robert Drew Hicks, Loeb Classical Library, Cambridge, MA, 1925.

Drews, C & W Han, "Dynamics of wind setdown at Suez and the eastern Nile Delta," *PLoS ONE*, 5:8, 2010, e12481, DOI:10.1371/ journal.pone.0012481.

Dundes, A (ed.), *Sacred Narrative: Readings in the theory of myth*, University of California Press, Berkeley, 1984.

Henry, J, Knowledge is Power, Icon Books, London, 2002.

Hyman, A & J Walsh, *Philosophy in the Middle Ages*, 2nd edn, Hackett, Indianapolis, 1973.

Katz, VJ, *A History of Mathematics: An introduction*, 2nd edn, Addison-Wesley, 1998.

Lopez, CA, "Franklin and Mesmer: an encounter," *The Yale Journal of Biology and Medicine*, 66:4, 1993, pp. 325–31.

Plato, *The Republic*, 2nd edn, Penguin Classics, London, 2007 [c. 360 BCE].

Radcliffe-Brown, AR, "The rainbow-serpent myth of Australia," *Journal of the Royal Anthropological Institute of Great Britain and Ireland*, 56, 1926, pp. 19–25.

Reale, G, *From the Origins to Socrates*, trans. JR Catan, SUNY Press, Albany, NY, 1987.

Rodriguez-Fernandez, JL, "Ockham's razor," *Endeavor*, 23:3, 1999, pp. 121–25.

Segal, R, *Myth: A very short introduction*, Oxford University Press, Oxford, 2004.

Steffens, B, *Ibn al-Haytham: First scientist*, Morgan Reynolds Publishing, Greensboro, 2006.

Whewell, W, *The Philosophy of the Inductive Sciences, founded upon their history*, JW Parker, London, 1840.

CHAPTER 3. THE PITIFUL MONSTER

Alexander, H, "Parents guilty of manslaughter over daughter's eczema death," *Sydney Morning Herald*, June 5, 2009.

BBC News, "Pope's condom stance sparks row," March 18, 2009, <news.bbc.co .uk/2/hi/7950671.stm>, accessed October 17, 2010.

Center for Disease Control, "Opportunistic infections and Kaposi's sarcoma among Haitians in the United States," *Morbidity and Mortality Weekly Report*, 31:353–4, 1982, pp. 360–1.

Chigwedere, P, GR Seage, S Gruskin, TH Lee & M Essex, "Lost benefits," *Journal of Acquired Immune Deficiency Syndrome*, 49:4, 2008, p. 410.

Christianson, GE, "Kepler's Somnium: Science Fiction and the Renaissance scientist," Science Fiction Studies, 3:1, 1976, <www.depauw.edu/sfs/back issues/8/christianson8art.htm>, accessed September 21, 2010.

Clarke, AC, *Profiles of the Future: An enquiry into the limits of the possible,* Henry Holt & Co., New York, 1973.

Committee for the Oversight of AIDS Activities, *Confronting AIDS: Update 1988*, Institute of Medicine of the US National Academy of Sciences, Washington, DC, 1998, p. 2.

Department of Innovation, Industry and Research, *Inspiring Australia: A national strategy for engagement with the sciences*, Commonwealth of Australia, Canberra, 2010.

Durant, J, GA Evans & GP Thomas, "The public understanding of science," *Nature*, 340, 1989, pp. 11–14.

Grmek, MD, RC Maulitz & J Duffin, *History of AIDS: Emergence and origin of a modern pandemic*, Princeton University Press, Princeton, 1993.

Hare, R, *Experimental Investigation of the Spirit Manifestations, Demonstrating the Existence of Spirits and Their Communion with Mortals*, Partridge & Brittan, New York, 1856.

Hilgartner, S, "The Sokal Affair in Context," *Science, Technology, & Human Values*, 22:4, 1997, pp. 506–22.

Hirshfeld, A, *The Electric Life of Michael Faraday*, Walker & Co., New York, 2006.

Ho, MW & J Cummins, "London drug trial catastrophe—collapse of science and ethics," Institute of Science in Society, 30, 2006, pp. 44–5 <www.isis .org.uk/LDTC.php>, accessed October 17, 2010.

James, F, *Guides to the Royal Institution of Great Britain: 1, History*, Royal Institution of Great Britain, London, 2000.

Jones, VA, "The white coat: why not follow suit?" *Journal of the American Medical Association*, 281:5, 1999, p. 478.

Knight, D, *Humphry Davy: Science and power*, Cambridge University Press, Cambridge, UK, 1992.

Layton, D, *Science for the People: The origins of the school science curriculum in England*, Allen & Unwin, London, 1973.

Mason, CL & JB Kahle, "Draw-a-scientist test: future implications," *School Science and Mathematics Association*, 91:5, 1991, pp. 193–8.

Mehta, A, "Birth of a drug," *Chemical & Engineering News*, March 22, 2004, pp. 51–6.

Papadopulos-Eleopulos, E, "A mitotic theory," *Journal of Theoretical Biology*, 96, 1982, pp. 741–58.

Potts, M, et al., "Reassessing HIV Prevention," *Science*, 320:5877, 2008, pp. 749–50.

Shelley, M, *Frankenstein, Penguin Books, London, 2003 [1818]*.

Smith, RA, *Encyclopedia of AIDS: A social, political, cultural, and scientific record of the HIV epidemic*, Taylor & Francis, Abingdon, UK, 1998.

Sokal, A, "Transgressing the boundaries: toward a transformative hermeneutics of quantum gravity," *Social Text*, 46, 1995, pp. 217–52.

Supreme Court of South Australia, *R v Parenzee*, [2007] SASC 143, April 27, 2007, <www.austlii.edu.au/au/cases/sa/SASC/2007/143. html>, accessed October 17, 2010.

Swinburne University of Technology, *The Swinburne National Technology and Society Monitor*, SPRU and CATI Executive Committee, Hawthorn, 2009.

Therapeutic Goods Association, *The Australian Clinical Trial Handbook*, Australian Government Department of Health and Ageing, Canberra, 2006.

Timbrell, J, *The Poison Paradox*, Oxford University Press, Oxford, 2005.

UNAIDS, *Report on the Global AIDS Epidemic*, UNAIDS, Geneva, 2008.

United States National Institute of Health, "Understanding clinical trials," 2007, <clinicaltrials.gov/ct2/info/understand>, accessed October 17, 2010.

United States National Science Foundation, Division of Science Resources Statistics, *NSF Survey of Public Attitudes Toward and Understanding of Science and Technology*, United States National Science Foundation, Washington, DC, 2001.

Voltaire, *Micromegas*, trans. P Phalen, Project Gutenberg, 1752.

Yao, L, "AMA weighs whether docs should hang up their white coats," *Wall Street Journal*, Health Blog, June 12, 2009.

CHAPTER 4. THE LOGICAL ALIEN

Dumitriu, A, *History of Logic*, Abacus Press, Tunbridge Wells, UK, 1977.

Gardner, M, *Fads, Facts and Fallacies in the Name of Science*, Dover Publications, New York, 1957.

Harris, S, "Toward a Science of Morality," *Huffington Post*, May 7, 2010, <www.huffingtonpost.com/sam-harris/a-science-of-morality_b_567185 .html>, accessed October 1, 2010.

Hume, D, *A Treatise of Human Nature*, NuVision Publications, Sioux Falls, SD, [1739].

Huxley, TH, *Evolution and Ethics*, Princeton University Press, Princeton, NJ, 2009 [1893].

Kahana, D, H Jenkins-Smith & D Braman, "Cultural cognition of scientific consensus," *Journal of Risk Research*, 2010, pp. 1–28, DOI: 10.1080/13669877 .2010.511246.

Kneale, W & M Kneale, *The Development of Logic,* Oxford University Press, Oxford, 1962.

Moore, GE, *Principia Ethica*, Cambridge University Press, Cambridge, UK, 1994 [1903].

Skinner, BF, "Superstition in the pigeon," *Journal of Experimental Psychology*, 38, 1947, pp.168–72.

Stanovich, K, *What Intelligence Tests Miss: The psychology of rational thought*, Yale University Press, London, 2009.

CHAPTER 5. THE CLEVER HORSE

Boroditsky, L, "How does language shape the way we think?," in M Brockman (ed.), *What's Next? Dispatches on the future of science*, Vintage Press, New York, 2009.

Chou, CW, DB Hume, T Rosenband & DJ Wineland, "Optical clocks and relativity," *Science*, 329, 2010, pp. 1630–3.

Dunbar, K, "How scientists really reason: scientific reasoning in real-world laboratories," in RJ Sternberg, & J Davidson (eds), *Mechanisms of Insight*, MIT Press, Cambridge, MA, 1995.

———. & J Fugelsang, "Causal thinking in science: how scientists and students interpret the unexpected," in ME Gorman, RD Tweney, D Gooding & A Kincannon (eds), *Scientific and Technical Thinking*, Lawrence Erlbaum Associates, Mahwah, NJ, 2005.

Henley, G & JE Harrison, *Injury Deaths, Australia 2004–05*, Injury research and statistics series, no. 51, Australian Institute of Health and Welfare, Canberra, 2009.

Heyn, ET, "Berlin's wonderful horse," *New York Times*, September 4, 1904.

Janes, LM, "Jeer pressure: the behavioral effects of observing ridicule of others," *Personality and Social Psychology Bulletin*, 26:4, 2000, pp. 474–85.

Kahneman, D & A Tversky, "The framing of decisions and the psychology of choice," *Science*, 211:4481, 1981, pp. 453–8.

———. 'Prospect theory: an analysis of decision under risk," *Econometrica*, 47:2, 1979, pp. 263–92.

Keeney, RL, "Personal decisions are the leading cause of death," *Operations Research*, 56, 2008, pp. 1335–47.

Kelly, I, J Rotton & R Culver, "The moon was full and nothing happened: a review of studies on the moon and human behavior," *Skeptical Inquirer*, 10:2, 1986, pp. 129–43.

Kensinger, E, "Negative emotion enhances memory accuracy: behavioral and neuroimaging evidence," *Current Directions in Psychological Science*, 16, 2007, pp. 213–18.

Kuhn, D & D Dean, "Metacognition: a bridge between cognitive psychology and educational practice," *Theory into Practice*, 43:4, 2004, pp. 268–73.

Loftus, EF & JC Palmer, "Reconstruction of automobile destruction: an example of the interaction between language and memory," *Journal of Learning and Verbal Behavior*, 13, 1974, pp. 585–9.

Mills, D, *The Encyclopedia of Applied Animal Behavior and Welfare*, Cambridge University Press, Cambridge, UK, 2010, pp. 106–7.

People for the Ethical Treatment of Animals, "Save the sea kittens," <features.peta.org/PETASeaKittens/>, accessed October 2, 2010.

Pinker, S, *The Language Instinct: How the mind creates language*, Harper Perennial Modern Classics, Sydney, 2000.

Sagan, C, *Cosmos*, PBS television series, Arlington, 1980, episode 12.

Singh, S, "Beware the spinal tap," *Guardian*, April 19, 2008, <www. guardian.co.uk/commentisfree/2008/apr/19/controversiesinscience-health>, accessed October 5, 2010.

Tversky, B & EJ Marsh, "Biased retellings of events yield biased memories," *Cognitive Psychology*, 40, 2000, pp. 1–38.

Wason, PC, "Reasoning," in B Foss (ed.), *New Horizons in Psychology*, Penguin Books, London, 1966.

Watts, P, "Because as we all know, the Green Party runs the world," November 22, 2009, <www.rifters.com/crawl/?p=886>, accessed 12 November 2010.

CHAPTER 6. THE SCIENCE GRAVEYARD

American Psychological Association, *Resolution on Facilitated Communication by the American Psychological Association*, adopted in Council, APA, Los Angeles, 1994, <www.apa.org/divisions/ div33/fcpolicy.html>, accessed October 10, 2010.

Autism National Committee, *Autism National Committee (Autcom) Policy And Principles Regarding Facilitated Communication*, Autism National Committee, Forest Knolls, CA, 2008, <www.autcom. org/articles/PPFC.pdf>, accessed October 12, 2010.

Brain Gym/Educational Kinesiology Australia, <www.braingym. org.au>, accessed October 12, 2010.

Cook, T, *Samuel Hahnemann—The founder of homeopathy*, Thorsons, London, 1981.

Crossley, R, *Annie's Coming Out*, Penguin Books, Ringwood, 1984.

Dally, A, *Fantasy Surgery: 1880–1930*, Rodopi, New York, 1996.

Dinsdale, T, *Loch Ness Monster*, 4th edn, Routledge, Abingdon, UK, 1982.

Eason, C, *Fabulous Creatures, Mythical Monsters, and Animal Power Symbols: A handbook*, Greenwood Publishing Group, Santa Barbara, CA, 2007, p. 143.

Emmer, R, *Loch Ness Monster: Fact or fiction?*, Infobase Publishing, New York, 2010.

Geller, U, *Uri Geller's Little Book of Mind-Power: Maximize your will to win*, Robson Books, London, 1998.

Goode, E, *Paranormal Beliefs: A sociological introduction*, Waveland Press, 2000.

Gould, SJ, *Rocks of Ages: Science and religion in the fullness of life*, Ballantine Books, New York, 2002.

Gratzer, W, *The Undergrowth of Science*, Oxford University Press, Oxford, 2000.

Haas, M, R Cooperstein & D Peterson, "Disentangling manual muscle testing and Applied Kinesiology: critique and reinterpretation of a literature review," *Chiropractic & Osteopathy*, 15:11, 2007, DOI:10.1186/1746-1340-15-11.

Hobhouse, RW, *Life of Christian Samuel Hahnemann*, B. Jain Publishers, New Delhi, 2001 [1921].

Hyman, R, "Psychology and 'Alternative Medicine': the mischief-making of ideomotor action," *The Scientific Review of Alternative Medicine*, 3:2, 1999, <www.sram.org/0302/ideomotor.html>, accessed November 17, 2010.

Kenney, JJ, R Clemens & KD Forsythe, "Applied kinesiology unreliable for assessing nutrient status," *Journal of the American Dietetic Association*, 88, 1998, pp. 698–704.

Lienesch, M, *In the Beginning: Fundamentalism, the Scopes trial, and the making of the antievolution movement*, University of North Carolina Press, Chapel Hill, NC, 2007.

Macpherson, E, W Jones & M Segonzac, "A new squat lobster family of *Galatheoidea* (*Crustacea, Decapoda, Anomura*) from the hydrothermal vents of the Pacific-Antarctic Ridge," *Zoosystema*, 27:4, 2006, pp. 709–23.

Maddox, J, J Randi & W Stewart, "'High-dilution' experiments a delusion," *Nature*, 334:6180, 1988, pp. 287–90.

Singh, S & E Ernst, *Trick or Treatment: Alternative medicine on trial*, WW Norton & Co. Inc., New York, 2009.

World Health Organization, *Acupuncture: Review and Analysis of Reports on Controlled Clinical Trials*, WHO, Geneva, 2003.

Young, A, "Some implications of medical beliefs and practices for Social Anthropology," *American Anthropologist*, new series, 78:1, 1976, pp. 5–24.

CHAPTER 7. THE TANGLED WEB

Abbate, J, *Inventing the Internet*, MIT Press, Cambridge, MA, 1999.

Asaria, P & E MacMahon, "Measles in the United Kingdom: can we eradicate it by 2010?," *British Medical Journal*, 333:7574, 2006, pp. 890–5.

Feldhay, R, *Galileo and the Church: Political inquisition or critical dialogue?* Cambridge University Press, Cambridge, UK, 1995.

Finocchiaro, MA, *The Galileo Affair: A documentary history*, University of California Press, Berkeley, CA, 1989.

———. *Defending Copernicus and Galileo: Critical reasoning in the two affairs*, Springer, New York, 2010.

Murch, S, "Separating inflammation from speculation in autism," *Lancet*, 362:9394, 2003, pp. 1498–9.

Nachman, RG, "Positivism and revolution in Brazil's first republic: the 1904 revolt," *The Americas*, 34:1, 1977, pp. 20–39.

Plato, *Phaedrus*, Forgotten Books, 1988 [c. 360 BCE].

Russell, Sir Muir, *The Independent Climate Change E-mails Review*, July 2010, <www.cce-review.org>, accessed October 17, 2010.

Technorati.com, "State of the Blogosphere 2010," <technorati.com/blogging/article/state-of-the-blogosphere-2010-introduction/>, accessed November 12, 2010.

Triggle, N, "MMR scare doctor 'acted unethically.' panel finds," BBC News, January 28, 2010, <news.bbc.co.uk/2/hi/health/8483865. stm>, accessed November 17, 2010.

Wakefield, AJ, et al., "Ileal-lymphoid-nodular hyperplasia, nonspecific colitis, and pervasive developmental disorder in children," *Lancet*, 352:9123, 1998, pp. 234–5.

Wolfe, R & L Sharp, "Anti-vaccinationists past and present," *British Medical Journal*, 325, 2002, pp. 430–2.

CHAPTER 8. THE PROGRESSIVE HUMAN

Blackmore, S & T Troscianko, "Belief in the paranormal: probability judgments, illusory control, and the chance baseline shift," *British Journal of Psychology*, 76, 1985, p. 459.

Bridgstock, M, *Beyond Belief*, Cambridge University Press, Cambridge, UK, 2009.

Briggs, A & P Burke, *A Social History of the Media: From Gutenberg to the internet*, 2nd edn, Cambridge, UK, 2005.

Cannell, JC, *The Secrets of Houdini*, Courier Dover Publications, New York, 1973.

Doyle, AC, *A Study in Scarlet*, Penguin Books, London, 1982 [1887].

Gal, D & Rucker D, "When in doubt, shout! paradoxical influences of doubt on proselytizing," *Psychological Science*, October 13, 2010, DOI: 10.1177/0956797610385953.

Kruger, J & D Dunning, "Unskilled and unaware of it: how difficulties in recognizing one's own incompetence lead to inflated self-assessments," *Journal of Personality and Social Psychology*, 77:6, 1999, pp. 1121–34.

Nordland, R, "Iraq swears by bomb detector US sees as useless," *The New York Times*, November 3, 2009.

Pelli, DG & C Bigelow, "A writing revolution," *Seed magazine*, October 20, 2009, <seedmagazine.com/content/article/a_writing_revolution>,accessed October 24, 2010.

Priest, P, "The effectiveness of instruction in scientific reasoning in altering paranormal beliefs in high school students," PhD dissertation, La Sierra University, Riverside, CA, 1995.

Russell, B, *Mortals and Others: Bertrand Russell's American essays, 1931–1935*, Routledge, Abingdon, UK, 1998.

INDEX

239